# Poultry Science

# Poultry Science

Edited by
**Gideon Waddell**

Larsen & Keller
www.larsen-keller.com

Poultry Science
Edited by Gideon Waddell
ISBN: 978-1-63549-229-3 (Hardback)



Larsen & Keller
Published by Larsen and Keller Education,
5 Penn Plaza,
19th Floor,
New York, NY 10001, USA

**Cataloging-in-Publication Data**

Poultry science / edited by Gideon Waddell.
p. cm.
Includes bibliographical references and index.
ISBN 978-1-63549-229-3
1. Poultry. 2. Poultry farms. 3. Poultry products. 4. Poultry--Diseases. I. Waddell, Gideon.
SF487 .P68 2017
636.5--dc23



Printed and bound in China.

For more information regarding Larsen and Keller Education and its products, please visit the publisher's website www.larsen-keller.com

# Table of Contents

# Preface

This book is compiled in such a manner, that it will provide in-depth knowledge about the theory and practice of poultry science. The main aim of the text is to familiarize students with the fundamental concepts of poultry science and make them understand the primary techniques used in this field. Poultry science is the field of agriculture which deals with the raising and management of poultry birds like geese, chicken, turkey and ducks, for food. The topics covered in this extensive text deal with the core subjects of poultry science. This textbook aims to serve as a resource guide for students and contribute to the growth of the discipline. The various sub-fields of poultry science along with technological progress that have future implications are glanced at in this book.

A short introduction to every chapter is written below to provide an overview of the content of the book:

**Chapter 1** - Birds that are kept for the purpose of producing eggs, meat and feathers are known as poultry. Poultry has existed for the past several thousand years. Some of the birds used for poultry are chicken, duck, emu, goose and ostrich. This chapter will provide an integrated understanding of poultry; **Chapter 2** - Farms which raise birds such as chickens, ducks and turkeys for the purpose of producing meat and eggs is termed as poultry farming. It can be alternative or intensive according to the procedures followed. This text is an overview of the subject matter incorporating all the major aspects of poultry farming; **Chapter 3** - The essential concepts of poultry science are capon, feed conversion ratio, broodiness, plumping, organic egg production, poultry show etc. A capon is a rooster which has been castrated whereas plumping is the process by which some companies inject raw chicken meat with saltwater or with chicken stock. The following section provides a glimpse of the essential concepts in poultry science; **Chapter 4** - The methods and techniques discussed within this chapter are free range, yarding, battery cage, furnished cages, debeaking, forced molting etc. Animals that can roam around free for some part of the day are denoted by the word free range where as yarding is the custom of fencing animals in order to keep them safe. Yarding is a method of intensive grazing. This text discusses the methods and techniques of poultry farming in a critical manner providing key analysis to the subject manner; **Chapter 5** - Various breeds of livestock are used in poultry. Some of these birds are Andalusian chicken, Bresse Gauloise, Delaware chicken, Asil chicken, broiler and Indian Runner ducks. This section has been carefully written to provide an easy understanding of the numerous kinds of livestock in poultry; **Chapter 6** - Poultry products are used mainly for food. Some of the chapters included in this section are chicken as food, duck as food, feather meal, egg as food and down feather. The chapter closely examines all the poultry products that are used in day-to-day life; **Chapter 7** - Aspergillosis is a group of diseases which is caused by a fungus named aspergillus. It causes diseases such as tuberculosis or chronic obstructive pulmonary disease. A number of diseases have been mentioned in this chapter, some of these are avian infectious bronchitis, botulism, infectious bursal disease and marek's disease. The subject matter of this text is very important as poultry can cause disease that can cause death and infections; **Chapter 8 -** The various poultry standards elucidated in this chapter are American standard of perfection, Australian poultry standards and British poultry standard. These ensure that efficient and effective food safety practices are being maintained. The aspects elucidated in this section are of vital importance, and provides a better understanding of poultry science.

I extend my sincere thanks to the publisher for considering me worthy of this task. Finally, I thank my family for being a source of support and help.

**Editor**

# 1

# Introduction to Poultry

Birds that are kept for the purpose of producing eggs, meat and feathers are known as poultry. Poultry has existed for the past several thousand years. Some of the birds used for poultry are chicken, duck, emu, goose and ostrich. This chapter will provide an integrated understanding of poultry.

## Poultry

Poultry of the World

Poultry are domesticated birds kept by humans for the eggs they produce, their meat, their feathers, or sometimes as pets. These birds are most typically members of the superorder Galloanserae (fowl), especially the order Galliformes (which includes chickens, quails and turkeys) and the family Anatidae, in order Anseriformes, commonly known as "waterfowl" and including domestic ducks and domestic geese. Poultry also includes other birds that are killed for their meat, such as the young of pigeons (known as squabs) but does not include similar wild birds hunted for sport or food and known as game. The word "poultry" comes from the French/Norman word *poule*, itself derived from the Latin word *pullus*, which means small animal.

The domestication of poultry took place several thousand years ago. This may have originally been as a result of people hatching and rearing young birds from eggs collected from the wild, but later involved keeping the birds permanently in captivity. Domesticated chickens may have been used for cockfighting at first and quail kept for their songs, but soon it was realised how useful it was having a captive-bred source of food. Selective breeding for fast growth, egg-laying ability, conformation, plumage and docility took place over the centuries, and modern breeds often look very different from their wild ancestors. Although some birds are still kept in small flocks in extensive systems, most birds available in the market today are reared in intensive commercial enterprises. Poultry is the second most widely eaten type of meat globally and, along with eggs, provides nutritionally beneficial food containing high-quality protein accompanied by a low proportion of fat. All poultry meat should be properly handled and sufficiently cooked in order to reduce the risk of food poisoning.

## Definition

The word "poultry" comes from the Middle English "pultrie", from Old French *pouletrie*, from *pouletier*, poultry dealer, from *poulet*, pullet. The word "pullet" itself comes from Middle English *pulet*, from Old French *polet*, both from Latin *pullus*, a young fowl, young animal or chicken. The word "fowl" is of Germanic origin (cf. Old English *Fugol*, German *Vogel*, Danish *Fugl*).

"Poultry" is a term used for any kind of domesticated bird, captive-raised for its utility, and traditionally the word has been used to refer to wildfowl (Galliformes) and waterfowl (Anseriformes). "Poultry" can be defined as domestic fowls, including chickens, turkeys, geese and ducks, raised for the production of meat or eggs and the word is also used for the flesh of these birds used as food. The Encyclopædia Britannica lists the same bird groups but also includes guinea fowl and squabs (young pigeons). In R. D. Crawford's *Poultry breeding and genetics*, squabs are omitted but Japanese quail and common pheasant are added to the list, the latter frequently being bred in captivity and released into the wild. In his 1848 classic book on poultry, *Ornamental and Domestic Poultry: Their History, and Management*, Edmund Dixon included chapters on the peafowl, guinea fowl, mute swan, turkey, various types of geese, the muscovy duck, other ducks and all types of chickens including bantams. In colloquial speech, the term "fowl" is often used near-synonymously with "domesticated chicken" (*Gallus gallus*), or with "poultry" or even just "bird", and many languages do not distinguish between "poultry" and "fowl". Both words are also used for the flesh of these birds. Poultry can be distinguished from "game", defined as wild birds or mammals hunted for food or sport, a word also used to describe the flesh of these when eaten.

## Examples

| Bird | Wild ancestor | Domestication | Utilization | Picture |
|---|---|---|---|---|
| Chicken | Red junglefowl | Southeast Asia | meat, feathers, eggs, ornamentation, leather |  |
| Duck | Muscovy duck/ Mallard | various | meat, feathers, eggs |  |
| Emu | Emu | various, 20th century | meat, leather, oil |  |
| Egyptian goose | Egyptian goose | Egypt | meat, feathers, eggs, ornamentation |  |
| Goose | Greylag goose/ Swan goose | various | meat, feathers, eggs |  |
| Indian peafowl | Indian Peafowl | various | meat, feathers, ornamentation, landscaping |  |
| Mute swan | Mute swan | various | feathers, eggs, land-scaping |  |
| Ostrich | Ostrich | various, 20th century | meat, eggs, feathers, leather |  |

| | | | | |
|---|---|---|---|---|
| Pigeon | Rock dove | Middle East | meat, feathers, ornamentation |  |
| Quail | Japanese quail, Northern bob-white | Japan, Virginia | meat, eggs, feathers, pets |  |
| Turkey | Wild turkey | Mexico | meat, feathers |  |
| Grey francolin | Grey francolin | Pakistan, North India | meat, fighting, pets |  |
| Guinea-fowl | Helmeted guineafowl | Africa | meat, pest consumption, and alarm calling |  |
| Common pheasant | Common pheasant | Eurasia | meat |  |
| Golden pheasant | Golden pheasant | Eurasia | meat, mainly ornamental |  |
| Greater rhea | Greater Rhea | various, 20th century | meat, leather, oil, eggs |  |

## Chickens

Cock with comb and wattles

Chickens are medium-sized, chunky birds with an upright stance and characterised by fleshy red combs and wattles on their heads. Males, known as cocks, are usually larger, more boldly coloured, and have more exaggerated plumage than females (hens). Chickens are gregarious, omnivorous, ground-dwelling birds that in their natural surroundings search among the leaf litter for seeds, invertebrates, and other small animals. They seldom fly except as a result of perceived danger, preferring to run into the undergrowth if approached. Today's domestic chicken (*Gallus gallus domesticus*) is mainly descended from the wild red junglefowl of Asia, with some additional input from grey junglefowl. Domestication is believed to have taken place between 7,000 and 10,000 years ago, and what are thought to be fossilized chicken bones have been found in northeastern China dated to around 5,400 BC. Archaeologists believe domestication was originally for the purpose of cockfighting, the male bird being a doughty fighter. By 4,000 years ago, chickens seem to have reached the Indus Valley and 250 years later, they arrived in Egypt. They were still used for fighting and were regarded as symbols of fertility. The Romans used them in divination, and the Egyptians made a breakthrough when they learned the difficult technique of artificial incubation. Since then, the keeping of chickens has spread around the world for the production of food with the domestic fowl being a valuable source of both eggs and meat.

Since their domestication, a large number of breeds of chickens have been established, but with the exception of the white Leghorn, most commercial birds are of hybrid origin. In about 1800, chickens began to be kept on a larger scale, and modern high-output poultry farms were present in the United Kingdom from around 1920 and became established in the United States soon after the Second World War. By the mid-20th century, the poultry meat-producing industry was of greater importance than the egg-laying industry. Poultry breeding has produced breeds and strains to fulfil different needs; light-framed, egg-laying birds that can produce 300 eggs a year; fast-growing, fleshy birds destined for consumption at a young age, and utility birds which produce both an acceptable number of eggs and a well-fleshed carcase. Male birds are unwanted in the egg-laying industry and can often be identified as soon as they are hatch for subsequent culling. In meat breeds, these birds are sometimes castrated (often chemically) to prevent aggression. The resulting bird, called a capon, has more tender and flavorful meat, as well.

Roman mosaic depicting a cockfight

A bantam is a small variety of domestic chicken, either a miniature version of a member of a standard breed, or a “true bantam” with no larger counterpart. The name derives from the town of Bantam in Java where European sailors bought the local small chickens for their shipboard supplies. Bantams may be a quarter to a third of the size of standard birds and lay similarly small eggs. They are kept by small-holders and hobbyists for egg production, use as broody hens, ornamental purposes, and showing.

## Cockfighting

Cockfighting is said to be the world’s oldest spectator sport and may have originated in Persia 6,000 years ago. Two mature males (cocks or roosters) are set to fight each other, and will do so with great vigour until one is critically injured or killed. Breeds such as the Aseel were developed in the Indian subcontinent for their aggressive behaviour. The sport formed part of the culture of the ancient Indians, Chinese, Greeks, and Romans, and large sums were won or lost depending on the outcome of an encounter. Cockfighting has been banned in many countries during the last century on the grounds of cruelty to animals.

## Ducks

Ducks are medium-sized aquatic birds with broad bills, eyes on the side of the head, fairly long necks, short legs set far back on the body, and webbed feet. Males, known as drakes, are often larger than females (simply known as ducks) and are differently coloured in some breeds. Domestic ducks are omnivores, eating a variety of animal and plant materials such as aquatic insects, molluscs, worms, small amphibians, waterweeds, and grasses. They feed in shallow water by dabbling, with their heads underwater and their tails upended. Most domestic ducks are too heavy to fly, and they are social birds, preferring to live and move around together in groups. They keep their plumage waterproof by preening, a process that spreads the secretions of the preen gland over their feathers.

Pekin ducks

Clay models of ducks found in China dating back to 4000 BC may indicate the domestication of ducks took place there during the Yangshao culture. Even if this is not the case, domestication of the duck took place in the Far East at least 1500 years earlier than in the West. Lucius Columella, writing in the first century BC, advised those who sought to rear ducks to collect wildfowl eggs and put them under a broody hen, because when raised in this way, the ducks "lay aside their wild nature and without hesitation breed when shut up in the bird pen". Despite this, ducks did not appear in agricultural texts in Western Europe until about 810 AD, when they began to be mentioned alongside geese, chickens, and peafowl as being used for rental payments made by tenants to landowners.

It is widely agreed that the mallard (*Anas platyrhynchos*) is the ancestor of all breeds of domestic duck (with the exception of the Muscovy duck (*Cairina moschata*), which is not closely related to other ducks). Ducks are farmed mainly for their meat, eggs, and down. As is the case with chickens, various breeds have been developed, selected for egg-laying ability, fast growth, and a well-covered carcase. The most common commercial breed in the United Kingdom and the United States is the Pekin duck, which can lay 200 eggs a year and can reach a weight of 3.5 kg (7.7 lb) in 44 days. In the Western world, ducks are not as popular as chickens, because the latter produce larger quantities of white, lean meat and are easier to keep intensively, making the price of chicken meat lower than that of duck meat. While popular in *haute cuisine*, duck appears less frequently in the mass-market food industry. However, things are different in the East. Ducks are more popular there than chickens and are mostly still herded in the traditional way and selected for their ability to find sufficient food in harvested rice fields and other wet environments.

## Geese

An Emden goose, a descendent of the wild greylag goose

The greylag goose (*Anser anser*) was domesticated by the Egyptians at least 3000 years ago, and a different wild species, the swan goose (*Anser cygnoides*), domesticated in Siberia about a thousand years later, is known as a Chinese goose. The two hybridise with each other and the large knob at the base of the beak, a noticeable feature of the Chinese goose, is present to a varying extent in these hybrids. The hybrids are fertile and have resulted in several of the modern breeds. Despite their early domestication, geese have never gained the commercial importance of chickens and ducks.

Domestic geese are much larger than their wild counterparts and tend to have thick necks, an upright posture, and large bodies with broad rear ends. The greylag-derived birds are large and fleshy and used for meat, while the Chinese geese have smaller frames and are mainly used for egg production. The fine down of both is valued for use in pillows and padded garments. They forage on grass and weeds, supplementing this with small invertebrates, and one of the attractions of rearing geese is their ability to grow and thrive on a grass-based system. They are very gregarious and have good memories and can be allowed to roam widely in the knowledge that they will return home by dusk. The Chinese goose is more aggressive and noisy than other geese and can be used as a guard animal to warn of intruders. The flesh of meat geese is dark-coloured and high in protein, but they deposit fat subcutaneously, although this fat contains mostly monounsaturated fatty acids. The birds are killed either around 10 or about 24 weeks. Between these ages, problems with dressing the carcase occur because of the presence of developing pin feathers.

In some countries, geese and ducks are force-fed to produce livers with an exceptionally high fat content for the production of *foie gras*. Over 75% of world production of this product occurs in France, with lesser industries in Hungary and Bulgaria and a growing production in China. *Foie gras* is considered a luxury in many parts of the world, but the process of feeding the birds in this way is banned in many countries on animal welfare grounds.

## Turkeys

Male domesticated turkey sexually displaying by showing the snood hanging over the beak, the caruncles hanging from the throat, and the 'beard' of small, black, stiff feathers on the chest

Turkeys are large birds, their nearest relatives being the pheasant and the guineafowl. Males are larger than females and have spreading, fan-shaped tails and distinctive, fleshy wattles, called a snood, that hang from the top of the beak and are used in courtship display. Wild turkeys can fly, but seldom do so, preferring to run with a long, stratling gait. They roost in trees and forage on the ground, feeding on seeds, nuts, berries, grass, foliage, invertebrates, lizards, and small snakes.

The modern domesticated turkey is descended from one of six subspecies of wild turkey (*Meleagris gallopavo*) found in the present Mexican states of Jalisco, Guerrero and Veracruz. Pre-Aztec tribes in south-central Mexico first domesticated the bird around 800 BC, and Pueblo Indians inhabiting the Colorado Plateau in the United States did likewise around 200 BC. They used the feathers for robes, blankets, and ceremonial purposes. More than 1,000 years later, they became an important food source. The first Europeans to encounter the bird misidentified it as a guineafowl, a bird known as a "turkey fowl" at that time because it had been introduced into Europe via Turkey.

Commercial turkeys are usually reared indoors under controlled conditions. These are often large buildings, purpose-built to provide ventilation and low light intensities (this reduces the birds' activity and thereby increases the rate of weight gain). The lights can be switched on for 24-hrs/day, or a range of step-wise light regimens to encourage the birds to feed often and therefore grow rapidly. Females achieve slaughter weight at about 15 weeks of age and males at about 19. Mature commercial birds may be twice as heavy as their wild counterparts. Many different breeds have been developed, but the majority of commercial birds are white, as this improves the appearance of the dressed carcass, the pin feathers being less visible. Turkeys were at one time mainly consumed on special occasions such as Christmas (10 million birds in the United Kingdom) or Thanksgiving (60 million birds in the United States). However, they are increasingly becoming part of the everyday diet in many parts of the world.

## Quail

Japanese quail

The quail is a small to medium-sized, cryptically coloured bird. In its natural environment, it is found in bushy places, in rough grassland, among agricultural crops, and in other places with dense cover. It feeds on seeds, insects, and other small invertebrates. Being a largely ground-dwelling, gregarious bird, domestication of the quail was not difficult, although many of its wild instincts are retained in captivity. It was known to the Egyptians long before the arrival of chickens and was depicted in hieroglyphs from 2575 BC. It migrated across Egypt in vast flocks and the birds could sometimes be picked up off the ground by hand. These were the common quail (*Coturnix coturnix*), but modern domesticated flocks are mostly of Japanese quail (*Coturnix japonica*) which was probably domesticated as early as the 11th century AD in Japan. They were originally kept as songbirds, and they are thought to have been regularly used in song contests.

In the early 20th century, Japanese breeders began to selectively breed for increased egg production. By 1940, the quail egg industry was flourishing, but the events of World War II led to the complete loss of quail lines bred for their song type, as well as almost all of those bred for egg production. After the war, the few surviving domesticated quail were used to rebuild the industry, and all current commercial and laboratory lines are considered to have originated from this population. Modern birds can lay upward of 300 eggs a year and countries such as Japan, India, China, Italy, Russia, and the United States have established commercial Japanese quail farming industries. Japanese quail are also used in biomedical research in fields such as genetics, embryology, nutrition, physiology, pathology, and toxicity studies. These quail are closely related to the common quail, and many young hybrid birds are released into the wild each year to replenish dwindling wild populations.

## Other Poultry

Guinea fowl originated in southern Africa, and the species most often kept as poultry is the helmeted guineafowl (*Numida meleagris*). It is a medium-sized grey or speckled bird with a small naked head with colourful wattles and a knob on top, and was domesticated by the time of the ancient Greeks and Romans. Guinea fowl are hardy, sociable birds that subsist mainly on insects, but also consume grasses and seeds. They will keep a vegetable garden clear of pests and will eat the ticks that carry Lyme disease. They happily roost in trees and give a loud vocal warning of the approach of predators. Their flesh and eggs can be eaten in the same way as chickens, young birds being ready for the table at the age of about four months.

A squab is the name given to the young of domestic pigeons that are destined for the table. Like other domesticated pigeons, birds used for this purpose are descended from the rock pigeon (*Columba livia*). Special utility breeds with desirable characteristics are used. Two eggs are laid and incubated for about 17 days. When they hatch, the squabs are fed by both parents on "pigeon's milk", a thick secretion high in protein produced by the crop. Squabs grow rapidly, but are slow to fledge and are ready to leave the nest at 26 to 30 days weighing about 500 g (18 oz). By this time, the adult pigeons will have laid and be incubating another pair of eggs and a prolific pair should produce two squabs every four weeks during a breeding season lasting several months.

## Poultry Farming

Free-range ducks in Hainan Province, China

Worldwide, more chickens are kept than any other type of poultry, with over 50 billion birds being raised each year as a source of meat and eggs. Traditionally, such birds would have been kept extensively in small flocks, foraging during the day and housed at night. This is still the case in developing countries, where the women often make important contributions to family livelihoods through keeping poultry. However, rising world populations and urbanization have led to the bulk of production being in larger, more intensive specialist units. These are often situated close to where the feed is grown or near to where the meat is needed, and result in cheap, safe food being made available for urban communities. Profitability of production depends very much on the price of feed, which has been rising. High feed costs could limit further development of poultry production.

In free-range husbandry, the birds can roam freely outdoors for at least part of the day. Often, this is in large enclosures, but the birds have access to natural conditions and can exhibit their normal behaviours. A more intensive system is yarding, in which the birds have access to a fenced yard and poultry house at a higher stocking rate. Poultry can also be kept in a barn system, with no access to the open air, but with the ability to move around freely inside the building. The most intensive system for egg-laying chickens is battery cages, often set in multiple tiers. In these, several birds share a small cage which restricts their ability to move around and behave in a normal manner. The eggs are laid on the floor of the cage and roll into troughs outside for ease of collection. Battery cages for hens have been illegal in the EU since January 1, 2012.

Chickens raised intensively for their meat are known as "broilers". Breeds have been developed that can grow to an acceptable carcass size (2 kg (4.4 lb)) in six weeks or less. Broilers grow so fast, their legs cannot always support their weight and their hearts and respiratory systems may not be able to supply enough oxygen to their developing muscles. Mortality rates at 1% are much higher than for less-intensively reared laying birds which take 18 weeks to reach similar weights. Processing the birds is done automatically with conveyor-belt efficiency. They are hung by their feet, stunned, killed, bled, scalded, plucked, have their heads and feet removed, eviscerated, washed, chilled, drained, weighed, and packed, all within the course of little over two hours.

Both intensive and free-range farming have animal welfare concerns. In intensive systems, cannibalism, feather pecking and vent pecking can be common, with some farmers using beak trimming as a preventative measure. Diseases can also be common and spread rapidly through the flock. In extensive systems, the birds are exposed to adverse weather conditions and are vulnerable to predators and disease-carrying wild birds. Barn systems have been found to have the worst bird welfare. In Southeast Asia, a lack of disease control in free-range farming has been associated with outbreaks of avian influenza.

## Poultry Shows

In many countries, national and regional poultry shows are held where enthusiasts exhibit their birds which are judged on certain phenotypical breed traits as specified by their respective breed standards. The idea of poultry exhibition may have originated after cockfighting was made illegal, as a way of maintaining a competitive element in poultry husbandry. Breed standards were drawn up for egg-laying, meat-type, and purely ornamental birds, aiming for uniformity. Sometimes, poultry shows are part of general livestock shows, and sometimes they are separate events such as the annual "National Championship Show" in the United Kingdom organised by the Poultry Club of Great Britain.

## Poultry as Food

### Trade

Chicken and duck eggs on sale in Hong Kong

Poultry is the second most widely eaten type of meat in the world, accounting for about 30% of total meat production worldwide compared to pork at 38%. Sixteen billion birds are raised annually for consumption, more than half of these in industrialised, factory-like production units. Global broiler meat production rose to 84.6 million tonnes in 2013. The largest producers were the United States (20%), China (16.6%), Brazil (15.1%) and the European Union (11.3%). There are two distinct models of production; the European Union supply chain model seeks to supply products which can be traced back to the farm of origin. This model faces the increasing costs of implementing additional food safety requirements, welfare issues and environmental regulations. In contrast, the United States model turns the product into a commodity.

World production of duck meat was about 4.2 million tonnes in 2011 with China producing two thirds of the total, some 1.7 billion birds. Other notable duck-producing countries in the Far East include Vietnam, Thailand, Malaysia, Myanmar, Indonesia and South Korea (12% in total). France (3.5%) is the largest producer in the West, followed by other EU nations (3%) and North America (1.7%). China was also by far the largest producer of goose and guinea fowl meat, with a 94% share of the 2.6 million tonne global market.

Global egg production was expected to reach 65.5 million tonnes in 2013, surpassing all previous years. Between 2000 and 2010, egg production was growing globally at around 2% per year, but since then growth has slowed down to nearer 1%.

### Cuts of Poultry

Poultry is available fresh or frozen, as whole birds or as joints (cuts), bone-in or deboned, seasoned in various ways, raw or ready cooked. The meatiest parts of a bird are the flight muscles on its chest, called "breast" meat, and the walking muscles on the legs, called the "thigh" and "drumstick". The wings are also eaten (Buffalo wings are a popular example in the United States) and may be split into three segments, the meatier "drumette", the "wingette" (also called the "flat"), and the wing tip (also called the "flapper"). In Japan, the wing is frequently separated, and these parts are referred to as 手羽元 (*teba-moto* "wing base") and 手羽先 (*teba-saki* "wing tip").

In the poultry pavilion of the Rungis International Market, France

Dark meat, which avian myologists refer to as "red muscle", is used for sustained activity—chiefly walking, in the case of a chicken. The dark colour comes from the protein myoglobin, which plays a key role in oxygen uptake and storage within cells. White muscle, in contrast, is suitable only for short bursts of activity such as, for chickens, flying. Thus, the chicken's leg and thigh meat are dark, while its breast meat (which makes up the primary flight muscles) is white. Other birds with breast muscle more suitable for sustained flight, such as ducks and geese, have red muscle (and therefore dark meat) throughout. Some cuts of meat including poultry expose the microscopic regular structure of intracellular muscle fibrils which can diffract light and produce iridescent colours, an optical phenomenon sometimes called structural colouration.

## Health and Disease (Humans)

Cuts from plucked chickens

Poultry meat and eggs provide nutritionally beneficial food containing protein of high quality. This is accompanied by low levels of fat which have a favourable mix of fatty acids. Chicken meat contains about two to three times as much polyunsaturated fat as most types of red meat when measured by weight. However, for boneless, skinless chicken breast, the amount is much lower. A 100-g serving of baked chicken breast contains 4 g of fat and 31 g of protein, compared to 10 g of fat and 27 g of protein for the same portion of broiled, lean skirt steak.

A 2011 study by the Translational Genomics Research Institute showed that 47% of the meat and poultry sold in United States grocery stores was contaminated with *Staphylococcus aureus*, and 52% of the bacteria concerned showed resistance to at least three groups of antibiotics. Thorough cooking of the product would kill these bacteria, but a risk of cross-contamination from improper handling of the raw product is still present. Also, some risk is present for consumers of poultry meat and eggs to bacterial infections such as *Salmonella* and *Campylobacter*. Poultry products may become contaminated by these bacteria during handling, processing, marketing, or storage, resulting in food-borne illness if the product is improperly cooked or handled.

In general, avian influenza is a disease of birds caused by bird-specific influenza A virus that is not normally transferred to people; however, people in contact with live poultry are at the greatest risk of becoming infected with the virus and this is of particular concern in areas such as Southeast Asia, where the disease is endemic in the wild bird population and domestic poultry can become infected. The virus possibly could mutate to become highly virulent and infectious in humans and cause an influenza pandemic.

Bacteria can be grown in the laboratory on nutrient culture media, but viruses need living cells in which to replicate. Many vaccines to infectious diseases can be grown in fertilised chicken eggs. Millions of eggs are used each year to generate the annual flu vaccine requirements, a complex process that takes about six months after the decision is made as to what strains of virus to include in the new vaccine. A problem with using eggs for this purpose is that people with egg allergies are unable to be immunised, but this disadvantage may be overcome as new techniques for cell-based rather than egg-based culture become available. Cell-based culture will also be useful in a pandemic when it may be difficult to acquire a sufficiently large quantity of suitable sterile, fertile eggs.

## References


- Cherry, Peter; Morris, T. R. (2008). Domestic Duck Production: Science and Practice. CABI. pp. 1–7. ISBN 978-1-84593-441-5.
- Smith, Andrew F. (2006). The Turkey: An American Story. University of Illinois Press. pp. 4–5, 17. ISBN 978-0-252-03163-2.
- Pond, Wilson, G.; Bell, Alan, W. (eds.) (2010). Turkeys: Behavior, Management and Well-Being. Marcell Dekker. pp. 847–849. ISBN 0-8247-5496-4.
- Adler, Jerry; Lawler, Andrew (June 1, 2012). "How the Chicken Conquered the World". Smithsonian Magazine. Retrieved April 14, 2014.
- Buckland, R.; Guy, G. "Origins and Breeds of Domestic Geese". FAO Agriculture Department. Retrieved February 17, 2014.
- Viegas, Jennifer (February 1, 2010). "Native Americans First Tamed Turkeys 2,000 Years Ago". Retrieved February 19, 2014.
- Jacob, Jacquie; Pescatore, Tony; Cantor, Austin. "Keeping Guinea fowl" (PDF). University of Kentucky. Retrieved March 9, 2014.

- Bolla, Gerry (April 1, 2007). "Primefacts: Squab raising" (PDF). NSW Department of Primary Industries. ISSN 1832-6668. Retrieved March 9, 2014.
- "USDA Livestock & Poultry: World Markets & Trade". The Poultry Site. April 30, 2013. Retrieved February 24, 2014.
- "USDA International Livestock & Poultry: World Duck, Goose and Guinea Fowl Meat Situation". The Poultry Site. December 19, 2013. Retrieved March 9, 2014.
- "Global Poultry Trends: World Egg Production Sets a Record Despite Slower Growth". The Poultry Site. January 16, 2013. Retrieved February 24, 2014.
- "Information on Avian Influenza". Seasonal Influenza (Flu). Centers for Disease Control and Prevention. Retrieved March 3, 2014.
- "The evolution, and revolution, of flu vaccines". FDA: Consumer updates. U.S. Food and Drug Administration. January 18, 2013. Retrieved March 6, 2014.
- "Feinberg School: Nutrition Fact Sheet: Lipids". Northwestern University. Archived from the original on July 20, 2011. Retrieved August 24, 2009.

2

# Poultry Farming: An Integrated Study

Farms which raise birds such as chickens, ducks and turkeys for the purpose of producing meat and eggs is termed as poultry farming. It can be alternative or intensive according to the procedures followed. This text is an overview of the subject matter incorporating all the major aspects of poultry farming.

## Poultry Farming

Poultry farming is the raising of domesticated birds such as chickens, ducks, turkeys and geese for the purpose of farming meat or eggs for food. Poultry are farmed in great numbers with chickens being the most numerous. More than 50 billion chickens are raised annually as a source of food, for both their meat and their eggs. Chickens raised for eggs are usually called layers while chickens raised for meat are often called broilers. In the US, the national organization overseeing poultry production is the Food and Drug Administration (FDA). In the UK, the national organisation is the Department for Environment, Food and Rural Affairs (Defra).

### Intensive and Alternative

According to the researchers and scientists, 74% of the world's poultry meat, and 68 percent of eggs are produced in ways that are described as 'intensive'. One alternative to intensive poultry farming is free-range farming using lower stocking densities. Poultry producers routinely use nationally approved medications, such as antibiotics, in feed or drinking water, to treat disease or to prevent disease outbreaks. Some FDA-approved medications are also approved for improved feed utilization.

### Egg-laying Chickens – Husbandry Systems

Commercial hens usually begin laying eggs at 16–20 weeks of age, although production gradually declines soon after from approximately 25 weeks of age. This means that in many countries, by approximately 72 weeks of age, flocks are considered economically unviable and are slaughtered after approximately 12 months of egg production, although chickens will naturally live for 6 or more years. In some countries, hens are force moulted to re-invigorate egg-laying.

Environmental conditions are often automatically controlled in egg-laying systems.

For example, the duration of the light phase is initially increased to prompt the beginning of egg-laying at 16–20 weeks of age and then mimics summer daylength which stimulates the hens to continue laying eggs all year round; normally, egg production occurs only in the warmer months. Some commercial breeds of hen can produce over 300 eggs a year.

## Free-range

Commercial free range hens

Free range chickens being fed outdoors

Free-range poultry farming allows chickens to roam freely for a period of the day, although they are usually confined in sheds at night to protect them from predators or kept indoors if the weather is particularly bad. In the UK, the Department for Environment, Food and Rural Affairs (Defra) states that a free-range chicken must have day-time access to open-air runs during at least half of its life. Unlike in the United States, this definition also applies to free-range egg laying hens. The European Union regulates marketing standards for egg farming which specifies a minimum condition for free-range eggs that "hens have continuous daytime access to open-air runs, except in the case of temporary restrictions imposed by veterinary authorities". The RSPCA "Welfare standards for laying hens and pullets" indicates that the stocking rate must not exceed 1,000 birds per hectare (10 $m^2$ per hen) of range available and a minimum area of overhead shade/shelter of 8 $m^2$ per 1,000 hens must be provided.

Free-range farming of egg-laying hens is increasing its share of the market. Defra figures indicate that 45% of eggs produced in the UK throughout 2010 were free-range, 5% were produced in barn systems and 50% from cages. This compares with 41% being free-range in 2009.

Suitable land requires adequate drainage to minimise worms and coccidial oocysts, suitable protection from prevailing winds, good ventilation, access and protection from predators. Excess heat, cold or damp can have a harmful effect on the animals and their productivity. Free-range farmers have less control than farmers using cages in what food their chickens eat, which can lead to unreliable productivity, though supplementary feeding reduces this uncertainty. In some farms, the manure from free-range poultry can be used to benefit crops.

The benefits of free-range poultry farming for laying hens include opportunities for natural behaviours such as pecking, scratching, foraging and exercise outdoors.

Both intensive and free-range farming have animal welfare concerns. Cannibalism, feather pecking and vent pecking can be common, prompting some farmers to use beak trimming as a preventative measure, although reducing stocking rates would eliminate these problems. Diseases can be common and the animals are vulnerable to predators. Barn systems have been found to have the worst bird welfare. In South-East Asia, a lack of disease control in free range farming has been associated with outbreaks of Avian influenza.

## Organic

In organic egg-laying systems, chickens are also free-range. Organic systems are based upon restrictions on the routine use of synthetic yolk colourants, in-feed or in-water medications, other food additives and synthetic amino acids, and a lower stocking density and smaller group sizes. The Soil Association standards used to certify organic flocks in the UK, indicate a maximum outdoors stocking density of 1,000 birds per hectare and a maximum of 2,000 hens in each poultry house. In the UK, organic laying hens are not routinely beak-trimmed.

## Yarding

While often confused with free-range farming, yarding is actually a separate method of poultry culture by which chickens and cows are raised together. The distinction is that free-range poultry are either totally unfenced, or the fence is so distant that it has little influence on their freedom of movement. Yarding is common technique used by small farms in the Northeastern US. The birds are released daily from hutches or coops. The hens usually lay eggs either on the floor of the coop or in baskets if provided by the farmer. This husbandry technique can be complicated if used with roosters, mostly because of aggressive behavior.

## Battery Cage

## Battery Cages for Layer Hens

Bank of cages

The majority of hens in many countries are reared in battery cages, although the European Union Council Directive 1999/74/EC has banned the conventional battery cage in EU states from January 2012. These are small cages, usually made of metal in modern systems, housing 3 to 8 hens. The walls are made of either solid metal or mesh, and the floor is sloped wire mesh to allow the faeces to drop through and eggs to roll onto an egg-collecting conveyor belt. Water is usually provided by overhead nipple systems, and food in a trough along the front of the cage replenished at regular intervals by a mechanical chain.

The cages are arranged in long rows as multiple tiers, often with cages back-to-back (hence the term 'battery cage'). Within a single shed, there may be several floors containing battery cages meaning that a single shed may contain many tens of thousands of hens. Light intensity is often kept low (e.g. 10 lux) to reduce feather pecking and vent pecking. Benefits of battery cages include easier care for the birds, floor eggs which are expensive to collect are eliminated, eggs are cleaner, capture at the end of lay is expedited, generally less feed is required to produce eggs, broodiness is eliminated, more hens may be housed in a given house floor space, internal parasites are more easily treated, and labor requirements are generally much reduced.

In farms using cages for egg production, there are more birds per unit area; this allows for greater productivity and lower food costs. Floor space ranges upwards from 300 $cm^2$ per hen. EU standards in 2003 called for at least 550 $cm^2$ per hen. In the US, the current recommendation by the United Egg Producers is 67 to 86 $in^2$ (430 to 560 $cm^2$) per bird. The space available to battery hens has often been described as less than the size of a piece of A4 paper. Animal welfare scientists have been critical of battery cages because they do not provide hens with sufficient space to stand, walk, flap their wings, perch, or make a nest, and it is widely considered that hens suffer through boredom and frustration through being unable to perform these behaviours. This can lead to a wide range of abnormal behaviours, some of which are injurious to the hens or their cagemates.

## Furnished Cage

In 1999, the European Union Council Directive 1999/74/EC banned conventional battery cages for laying hens throughout the European Union from January 1, 2012; they were banned previously in other countries including Switzerland. In response to these bans, development of prototype commercial furnished cage systems began in the 1980s. Furnished cages, sometimes called 'enriched' or 'modified' cages, are cages for egg laying hens which have been designed to overcome some of the welfare concerns of battery cages whilst retaining their economic and husbandry advantages, and also provide some of the welfare advantages of non-cage systems. Many design features of furnished cages have been incorporated because research in animal welfare science has shown them to be of benefit to the hens. In the UK, the Defra "Code for the Welfare of Laying Hens" states furnished cages should provide at least 750 $cm^2$ of cage area per hen, 600 $cm^2$ of which should be usable; the height of the cage other than that above the usable area should be at least 20 cm at every point and no cage should have a total area that is less than 2000 $cm^2$. In addition, furnished cages should provide a nest, litter such that pecking and scratching are possible, appropriate perches allowing at least 15 cm per hen, a claw-shortening device, and a feed trough which may be used without restriction providing 12 cm per hen.

Modern egg laying breeds often suffer from osteoporosis which results in the chicken's skeletal system being weakened. During egg production, large amounts of calcium are transferred from bones to create egg-shell. Although dietary calcium levels are adequate, absorption of dietary calcium is not always sufficient, given the intensity of production, to fully replenish bone calcium. This can lead to increases in bone breakages, particularly when the hens are being removed from cages at the end of laying.

Countries such as Austria, Belgium or Germany are planning to ban furnished cages until 2025 additionally to the already banned conventional cages.

## Meat-producing Chickens – Husbandry Systems

Broilers in a production house

### Indoor Broilers

Meat chickens, commonly called broilers, are floor-raised on litter such as wood shavings, peanut shells, and rice hulls, indoors in climate-controlled housing. Under modern farming methods, meat chickens reared indoors reach slaughter weight at 5 to 9 weeks of age. The first week of chickens life they can grow 300 percent of their body size, a nine-week-old chicken can average over 9 pounds in body weight. At nine weeks a hen will average around 7 pounds and a rooster will weigh around 12 pounds, having a nine-pound average.

Broilers are not raised in cages. They are raised in large, open structures known as grow out houses. A farmer receives the birds from the hatchery at one day old. A grow out consist of 5 to 9 weeks according on how big the kill plant wants the chickens to be. These houses are equipped with mechanical systems to deliver feed and water to the birds. They have ventilation systems and heaters that function as needed. The floor of the house is covered with bedding material consisting of wood chips, rice hulls, or peanut shells. In some cases they can be grown over dry litter or compost. Because dry bedding helps maintain flock health, most growout houses have enclosed watering systems ("nipple drinkers") which reduce spillage.

Keeping birds inside a house protects them from predators such as hawks and foxes. Some houses are equipped with curtain walls, which can be rolled up in good weather to admit natural light and fresh air. Most growout houses built in recent years feature "tunnel ventilation," in which a bank of fans draws fresh air through the house.

Traditionally, a flock of broilers consist of about 20,000 birds in a growout house that measures 400/500 feet long and 40/50 feet wide, thus providing about eight-tenths of a square foot per bird. The Council for Agricultural Science and Technology (CAST) states that the minimum space is one-half square foot per bird. More modern houses are often larger and contain more birds, but the floor space allotment still meets the needs of the birds. The larger the bird is grown the fewer chickens are put in each house, to give the bigger bird more space per square foot.

Because broilers are relatively young and have not reached sexual maturity, they exhibit very little aggressive conduct.

Chicken feed consists primarily of corn and soybean meal with the addition of essential vitamins and minerals. No hormones or steroids are allowed in raising chickens.

## Issues With Indoor Husbandry

In intensive broiler sheds, the air can become highly polluted with ammonia from the droppings. In this case a farmer must run more fans to bring in more clean fresh air. If not this can damage the chickens' eyes and respiratory systems and can cause painful burns on their legs (called hock burns) and blisters on their feet. Broilers bred for fast growth have a high rate of leg deformities because the large breast muscles causes distortions of the developing legs and pelvis, and the birds cannot support their increased body weight. In cases where the chickens become crippled and can't walk farmers have to go in and pull them out. Because they cannot move easily, the chickens are not able to adjust their environment to avoid heat, cold or dirt as they would in natural conditions. The added weight and overcrowding also puts a strain on their hearts and lungs and Ascites can develop. In the UK, up to 19 million broilers die in their sheds from heart failure each year. In the case of no ventilation due to power failure during a heat wave 20,000 chicken can die in a short period of time. In a good grow out a farmer should sell between 92 and 96 percent of their flock. With a 1.80 to a 2.0 feed conversion ratio. After the marking of birds the farmer must clean out and repair for another flock. A farmer should average 4 to 5 grow outs a year.

## Indoor With Higher Welfare

Chickens are kept indoors but with more space (around 12 to 14 birds per square metre). They have a richer environment for example with natural light or straw bales that encourage foraging and perching. The chickens grow more slowly and live for up to two weeks longer than intensively farmed birds. The benefits of higher welfare indoor systems are the reduced growth rate, less crowding and more opportunities for natural behaviour.

## Free-range Broilers

Free-range broilers are reared under similar conditions to free-range egg laying hens. The breeds grow more slowly than those used for indoor rearing and usually reach slaughter weight at approximately 8 weeks of age. In the EU, each chicken must have one square metre of outdoor space. The benefits of free-range poultry farming include opportunities for natural behaviours such as pecking, scratching, foraging and exercise outdoors. Because they grow slower and have opportunities for exercise, free-range broilers often have better leg and heart health.

Turkeys on pasture at an organic farm

## Organic Broilers

Organic broiler chickens are reared under similar conditions to free-range broilers but with restrictions on the routine use of in-feed or in-water medications, other food additives and synthetic amino acids. The breeds used are slower growing, more traditional breeds and typically reach slaughter weight at around 12 weeks of age. They have a larger space allowance outside (at least 2 square metres and sometimes up to 10 square metres per bird). The Soil Association standards indicate a maximum outdoors stocking density of 2,500 birds per hectare and a maximum of 1,000 broilers per poultry house.

## Issues

## Humane Treatment

Battery cages

Chickens transported in a truck.

Animal welfare groups have frequently criticized the poultry industry for engaging in practices which they believe to be inhumane. Many animal rights advocates object to killing chickens for food, the "factory farm conditions" under which they are raised, methods of transport, and slaughter. Compassion Over Killing and other groups have repeatedly conducted undercover investigations at chicken farms and slaughterhouses which they allege confirm their claims of cruelty.

Conditions in chicken farms may be unsanitary, allowing the proliferation of diseases such as salmonella, E. coli and campylobacter. Chickens may be raised in very low light intensities, sometimes total darkness, to reduce injurious pecking. Concerns have been raised that companies growing single varieties of birds for eggs or meat are increasing their susceptibility to disease. Rough handling, crowded transport during various weather conditions and the failure of existing stunning systems to render the birds unconscious before slaughter, have also been cited as welfare concerns.

A common practice among hatcheries for egg-laying hens is the culling of newly hatched male chicks since they do not lay eggs and do not grow fast enough to be profitable for meat. There are plans to more ethically destroy the eggs before the chicks are hatched by "in-ovo" sex determination.

## Beak Trimming

Laying hens are routinely beak-trimmed at 1 day of age to reduce the damaging effects of aggression, feather pecking and cannibalism. Scientific studies have shown that beak trimming is likely to cause both acute and chronic pain.

The beak is a complex, functional organ with an extensive nervous supply including nociceptors that sense pain and noxious stimuli. These would almost certainly be stimulated during beak trimming, indicating strongly that acute pain would be experienced. Behavioural evidence of pain after beak trimming in layer hen chicks has been based on the observed reduction in pecking behavior, reduced activity and social behavior, and increased sleep duration. Severe beak trimming, or beak trimming birds at an older age, may cause chronic pain. Following beak trimming of older or adult hens, the nociceptors in the beak stump show abnormal patterns of neural discharge, which indicate acute pain.

Neuromas, tangled masses of swollen regenerating axon sprouts, are found in the healed stumps of birds beak trimmed at 5 weeks of age or older and in severely beak trimmed birds. Neuromas have been associated with phantom pain in human amputees and have therefore been linked to chronic pain in beak trimmed birds. If beak trimming is severe because of improper procedure or done in older birds, the neuromas will persist which suggests that beak trimmed older birds experience chronic pain, although this has been debated.

Beak-trimmed chicks will initially peck less than non-trimmed chickens, which animal behavioralist Temple Grandin attributes to guarding against pain. The animal rights activist, Peter Singer, claims this procedure is bad because beaks are sensitive, and the usual practice of trimming them without anaesthesia is considered inhumane by some. Some within the chicken industry claim that beak-trimming is not painful whereas others argue that the procedure causes chronic pain and discomfort, and decreases the ability to eat or drink.

## Antibiotics

### Overview of Antibiotic Use in Poultry

Antibiotics have been used in poultry farming in mass quantities since 1951, when the Food and Drug Administration (FDA) approved their use. Three years prior to the FDA's approval, scientists were investigating a phenomena in which chickens who were rooting through bacteria-rich manure were displaying signs of greater health than those who did not. Through testing, it was discovered that chickens who were fed a variety of vitamin B12 manufactured with the residue of a certain antibiotic grew 50 percent faster than those chickens who were fed B12 manufactured from a different source. Further testing confirmed that use of antibiotics did improve the health of the chickens, resulting in the chickens laying more eggs and experiencing lower mortality rates and less illness. Upon this discovery, farmers transitioned from expensive animal proteins to comparatively inexpensive antibiotics and B12. Chickens were now reaching their market weight at a much faster rate and at a lower cost. With a growing population and greater demand on the farmers, antibiotics appeared to be an ideal and cost-effective way to increase the output of poultry. Since this discovery, antibiotics have been routinely used in poultry production, but more recently have been the topic of debate secondary to the fear of bacterial antibiotic resistance.

Emerging Threats: Antibiotic Resistance The Centers for Disease Control (CDC), has identified the emergence of antibiotic resistance as a national threat. The concern over antibiotic use in livestock arises from the necessity antibiotics have in keeping populations disease-free. As of 2016, over 70 percent of FDA approved antibiotics are utilized in modern, high production poultry farms to prevent, control, and treat disease. The FDA released a report in 2009 estimating that 29 million pounds of antibiotics had been used in livestock in that year alone. However, surveillance of consumer exposure to antibiotics through poultry consumption is limited. More specifically in 2012, the FDA speculated the most significant public health threat in regard to antimicrobial use in animals is the exposure of antimicrobial resistant bacteria to humans. These statements are challenged by the American meat industry lobbyists that antibiotics are used responsibly and judiciously in order to ensure effectiveness.

Consumer Health effects Consumers are exposed to antibiotic resistance through consumption of poultry products that have prior exposure to resistant strains. In poultry husbandry, the practice of using medically important antibiotics can select for resistant strains of bacteria, which are then transferred to consumers through poultry meat and eggs. The CDC acknowledges this transferal pathway in their 2013 report of Antibiotic Resistant Threats in the United States. The annual rate of foodborne illness in the United States is one in six. For the 48 million individuals affected, antibiotics play a critical role in thwarting mortality rates. In a literature review conducted by the Review of Antimicrobial Resistance 100 out of 139 studies found evidence of a link between antibiotic use in animals and antibiotic resistance in consumers.

When a gram-negative bacterial infection is suspected in a patient, one of the first-line options for treatment is in the fluoroquinolone family. This, along with penicillin, is one of the first families of antibiotics utilized in the broiler industry. If this first-line treatment is not successful, a stronger class of antibiotics is typically used, however, there is a limitation on how many classes are available, as well as which medications are available on hospital formularies. There is also more drug toxicity affiliated with second and third line antibiotic options. This is one example why it is critical to keep as many first line antibiotic options available for human use.

Other issues are associated with duration and complexity of infection. On average, treatment for non-resistant bacteria is administered 11.5 hours after diagnosis, and treatment for resistant bacteria is administered 72 hours after diagnosis. This is a reflection of the additional threat of prolonged incubation, leading to greater potential for systemic disease, with higher morbidity and mortality associated with opportunities for complications, and prolonged treatment time. For example, of the two million people affected by resistant infections a year, 23,000 will die. Severity in mortality is coupled when exposed to high risk populations such as immunocompromised and elderly individuals in hospital and nursing home settings.

## History of US Federal Policy on Antibiotic Use in Livestock

- 1940s – Beginning of utilization of antibiotics in livestock feed
- 1951 - Antibiotics first FDA approved for use in poultry. Approved uses included

production (growth enhancement), treatment, control, or prevention of animal disease. Antibiotics were also available for purchase over the counter at that time.

- 1970 - FDA task force publication proposes limitations of utilizing antibiotics in livestock feed that are also used in humans.
- 1975 - Secondary to this publication, drug sponsors are required to submit studies demonstrating the antibiotics did not harm human health
- 1976 - Stuart Levy study demonstrating tetracycline resistant E. coli moving to consumers
- 1977 - FDA proposal to remove penicillin and tetracycline in subtherapeutic doses, however, request by Congress for further studies to be conducted.
- 1980 - National Academy of Science recruited by the FDA to conduct further studies, specifically for penicillins and tetracyclines. Conclusion from these studies indicated no sufficient evidence to ban these antibiotics.
- 1980s-early 2000s - Further studies continued, supported by the FDA
- 2003 - FDA issued guidance to pharmaceuticals for an approval process utilizing new antibiotics in animal feed. For antibiotics already in use, the FDA would have to withdraw approval for each individual medication.
- 2005 - Enrofloxacin, an already utilized antibiotic, was removed from poultry production. This took 5 years to accomplish.
- 2010 - FDA first draft of "voluntary" limitations of medically important antibiotics in livestock, and requirement of veterinarian oversight, which would later become "Guidance for Industry #209."
- 2011 - FDA removed original request from 1977 to remove penicillins and tetracyclines in feed.
- 2012 - FDA finalized "Guidance for Industry #209," which was implemented under the Veterinary Feed Directives. These guidelines were issued to pharmaceuticals.
- 2013 - FDA issues "Guidance for Industry #213," which provided additional information to pharmaceuticals for recommendations from #209.
- 2014 - All 26 pharmaceutical companies producing antibiotics used in livestock feed agreed to the FDA guidelines in #213. Gave total of 3 years to make all recommended changes.

## Current Federal Regulators

National Antimicrobial Resistance Monitoring System's (NARMS) Enteric Bacteria program - Established in 1996, and represents a collaboration between the USDA, FDA, and CDC. Its purpose is to organize these organizations into a drug monitoring program for antibiotics utilized in animal feed withe goal of maintaining their medical efficacy. There are three branches which oversee humans, retail meat, and food animals.

- USDA - Operating under the Food Safety and Inspection Service (FSIS). Main role is in charge of testing imported and domestic meat for antimicrobial resistant bacteria. If a 'residue violation' found, they may condemn the product. Regardless, funding and resources are not available for outbreak investigations at farms or ranches.
- FDA - Operating under the Center for Veterinary Medicine (CVM). Works with CDC to monitor retail meat.
- CDC - Monitors human samples.

## Vertical Integration

This is the current business structure utilized almost universally in the broiler, or chicken bred for meat, industry. This also began in the 1940s when antibiotics began to be utilized in livestock feed. Perdue is credited as the pioneer of this structure. The basis is centralization of production. 'Integrators' control cost, policy, and are the decision makers of production. They decide feed formulations, choice of antibiotic administration, and cover those costs in addition to veterinary services. They also own the poultry that is grown. Farmers are labeled as 'Growers' or 'Operators'. They own the land and buildings where the poultry is grown, and are essentially caretakers for the poultry growth to the Integrators. The benefit for Growers in this business structure is they are guaranteed payment from the Integrators, which is compensated in weight gained by each flock. Due to this structure, about 90% of broilers are raised within 60 miles of the processing plant. Integrators are large poultry companies such as Perdue, Tyson, Pilgrim's Pride, Koch Foods, etc. There are about 20 of these companies in the U.S. that control 96% of all broilers produced in 2011.

## Regulatory Surveys

There are two main surveys distributed to farmers by the federal government to aid in various regulations of the agricultural industry. They are the Agricultural and Resource Management Survey (ARMS) and the National Animal Health Monitoring Survey (NAHMS).

Agricultural and Resource Management Survey (ARMS) - Ran by the USDA's Economic Research Service (ERS) and National Agricultural Statistics Service (NASS). The main focus is finances of farming, production practices, and resource use. Seventeen total states are sampled every 5–6 years per livestock type, with the most recent surveys distributed to broiler farmers in 2006 and 2011. There was one question about utilization of antibiotics in poultry food or water, excluding use for illness treatment.

## Antibiotic Resistant Outbreaks from Poultry Meat

In order to minimize and prevent any residues of antibiotics in chicken meat, any chickens given antibiotics are required to have a "withdrawal" period before they can be slaughtered. Samples of poultry at slaughter are randomly tested by the FSIS, and show a very low percentage of residue violations. Although violations are minimal, these small amounts of antibiotics have still contributed to antibiotic resistant outbreaks in the U.S. There are five infectious agents that account for 90% of foodborne related deaths. Three consistently found in poultry are: Salmonella, Campylobacter, and Escherichia coli.

- 2014: Outbreak of Salmonella in 634 people across 29 states (38% hospitalized) from eating chicken from Foster Farms that was sold at Costco. 44/68 tested isolates were resistant to at least 1 drug (65%), and 4 of 5 chicken samples tested were drug resistant (80%).

- 2015: Outbreak of Salmonella in 15 people in 7 states (4 hospitalized) from eating frozen stuffed chicken produced by Barber Foods.

## Limitations & Challenges

One obstacle to gathering more comprehensive data on the use of antibiotics in feed is the majority of the poultry industry utilizes vertical integration. As a consequence, farmers are often unaware of what components go into the feed, including whether or not antibiotics are used. Also in antibiotic usage in general, there are criteria to define bacterial resistance to specific antibiotics, however, there are no standards to divide the bacteria into resistant and susceptible categories based on antibiotics utilized.

The poultry industry also plays a large part in the United States economy, both in domestic purchasing and through international demand. The USDA reports that the U.S. is the "world's largest producer and second largest exporter of poultry meat." In 2010, the U.S. produced 36.9 billion pounds of broiler meat and exported 6.8 billion pounds of broiler meat. This equates to an estimated retail value of 45 billion dollars in 2010.

Both the agricultural and pharmaceutical industries have been lobbying against legislation that seeks to quell non-therapeutic antibiotic use in livestock since the first introduction of such legislation in Congress in the 1970's. Despite scientific evidence suggesting a strong association between antibiotic use in poultry and other livestock, agribusiness lobbies such as The National Chicken Council argue that there is not sufficient evidence to purport that there is a measurable impact to humans and shifts the blame of the problem of antibiotic resistance to overprescribing in the field of medicine.

With antibiotic restrictions, integrators will bare the immediate costs of these changes, and would likely result in modified finances and contracts with growers. Also, public health agencies may not have adequate scientific evidence for making appropriate decisions for better public health outcomes, secondary to lack of research funds. As a reference, the US spends about $101 billion per year for both governmental and biomedical industrial research, which is only 5% of total health expenditures.

## Solutions

Several policies have been proposed to improve data collection and transparency in livestock production. For example, the 2013 Delivering Antimicrobial Transparency in Animals (DATA) Act proposed the enactment of policies to acquire more accurate documentation of antibiotic use in growth promotion by farmers, drug manufacturers, and the FDA. Also, the Preservation of Antibiotics for Medical Treatment Act (PAMTA) was enacted to eliminate the use of medically important antibiotics in livestock. In 2015, the Preventing Antibiotic Resistance Act (PARA) was passed with two components: requirement of drug companies to provide evidence that antibiotics that are approved for use in poultry, and that meat production does not add to the growing threat of antibiotic resistance in humans. Antimicrobial Stewardship Programs (ASPs) serve as an example of systematic monitoring and analysis of data via interdisciplinary and multi-sectoral collaboration.

Performing quality improvement in the process of livestock production is another focus. Some alternative methods include “improving hygiene, using enzymes, probiotics, prebiotics, and acids to improve health and utilizing bacteriocins, antimicrobial peptides, and bacteriophages as substitutes for antibiotics.” Adaptations of methods by other countries is an addtional focus. For example, the use of antibiotics in feed was banned in Sweden in 1985 with no compensatory increase in antibiotic usage in other sectors of production, proving that a ban can be successfully administered without unintended impacts on other categories.

Major producers in the poultry industry have also begun to make strides towards change, largely due to public concern over the widespread use of antibiotics in poultry. Some producers have started eliminating the use of antibiotics in order to produce and market chickens that may legally be labeled “antibiotic free”. In 2007, Perdue began phasing out all medically important antibiotics from its feed and hatcheries and began selling poultry products labeled “no antibiotics ever” under the Harvestland brand. Consumer response was positive and in 2014 Perdue also began phasing out ionophores from its hatchery and began using the “antibiotic free” labels on its Harvestland, Simply Smart and Perfect Portions products.

## Impacts of Change

As Guidance for Industry #213 has been voluntarily accepted, it will be a violation of the Federal Food, Drug, and Cosmetic Act to use antibiotics in livestock production for non-therapeutic purposes. However, as there is now a requirement for veterinary oversight and approval for antibiotics use, there is leeway in the interpretation of non-therapeutic purposes dependent on the situation. For example, per the FDA, "a veterinarian may determine, based on the client's production practices and history, that weaned beef calves arriving at a feedlot in bad weather after a lengthy transport are at risk to develop bacterial respiratory infection. In this case, the veterinarian might choose to preventively treat these calves with an antimicrobial approved for prevention of that bacterial infection."

The FDA is not trying to regulate all antimicrobials at this time - only those antibiotics which are considered "medically important." For example, bacitracin, a common antibiotic found in over the counter antibiotic ointments, is not classified as "medically important." Also, ionophores, which are not apart of human medicine but given for improving the health of livestock, are also not included in this regulation.

## Arsenic

Poultry feed can also include roxarsone or nitarsone, arsenical antimicrobial drugs that also promote growth. Roxarsone was used as a broiler starter by about 70% of the broiler growers between 1995 and 2000. The drugs have generated controversy because it contains arsenic, which is highly toxic to humans. This arsenic could be transmitted through run-off from the poultry yards. A 2004 study by the U.S. magazine Consumer Reports reported "no detectable arsenic in our samples of muscle" but found "A few of our chicken-liver samples has an amount that according to EPA standards could cause neurological problems in a child who ate 2 ounces of cooked liver per week or in an adult who ate 5.5 ounces per week." The U.S. Food and Drug Administration (FDA), however, is the organization responsible for the regulation of foods in America, and all samples tested were "far less than the... amount allowed in a food product."

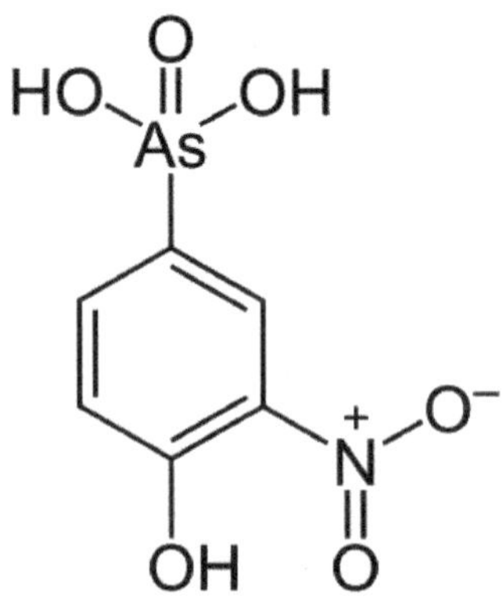


Roxarsone, a controversial arsenic compound used as a nutritional supplement for chickens.

## Growth Hormones

Hormone use in poultry production is illegal in the United States. Similarly, no chicken meat for sale in Australia is fed hormones. Several scientific studies have documented the fact that chickens grow rapidly because they are bred to do so, not because of growth hormones. A small producer of natural and organic chickens confirmed this assumption:

> "Using hormones to boost egg production was a brief fad in the Forties, but was abandoned because it didn't work. Using hormones to produce soft-meated roasters lasted into the Fifties, but the improved growth rates of normal, untreated broilers made the practice irrelevant—the broilers got as big as anyone wanted without chemicals. The only hormone that was ever used in any quantity on poultry (DES) was banned in 1959, and everyone but a few die-hard farmers had given up hormones by then, anyway. Hormones are now illegal in poultry and eggs."

## E. coli

According to Consumer Reports, "1.1 million or more Americans [are] sickened each year by undercooked, tainted chicken." A USDA study discovered *E. coli (Biotype I)* in 99% of supermarket chicken, the result of chicken butchering not being a sterile process. However, the same study also shows that the strain of *E. coli* found was always a non-lethal form, and no chicken had any of the pathenogenic O157:H7 serotype. Many of these chickens, furthermore, had relatively low levels of contamination.

Feces tend to leak from the carcass until the evisceration stage, and the evisceration stage itself gives an opportunity for the interior of the carcass to receive intestinal bacteria. (The skin of the carcass does as well, but the skin presents a better barrier to bacteria and reaches higher temperatures during cooking.) Before 1950, this was contained largely by not eviscerating the carcass at the time of butchering, deferring this until the time of retail sale or in the home. This gave the intestinal bacteria less opportunity to colonize the edible meat. The development of the "ready-to-cook broiler" in the 1950s added convenience while introducing risk, under the assumption that end-to-end refrigeration and thorough cooking would provide adequate protection. *E. coli* can be killed by proper cooking times, but there is still some risk associated with it, and its near-ubiquity in commercially farmed chicken is troubling to some. Irradiation has been proposed as a means of sterilizing chicken meat after butchering.

The aerobic bacteria found in poultry housing can include not only *E. coli*, but *Staphylococcus*, *Pseudomona*, *Micrococcus* and others as well. These contaminants can contribute to dust that often cause issues with the respiratory systems of both the poultry and humans working in the environment. If bacterial levels in the poultry drinking water reach high levels, it can result in bacterial diarrhoea which can lead to blood poisoning should the bacteria spread from the damaged intestines.

*Salmonella* too can be stressful on poultry production. How it causes disease has been investigated in some detail.

## Avian influenza

There is also a risk that crowded conditions in chicken farms will allow avian influenza (bird flu) to spread quickly. A United Nations press release states: "Governments, local authorities and international agencies need to take a greatly increased role in combating the role of factory-farming, commerce in live poultry, and wildlife markets which provide ideal conditions for the virus to spread and mutate into a more dangerous form..."

## Efficiency

Farming of chickens on an industrial scale relies largely on high protein feeds derived from soyabeans; in the European Union the soybean dominates the protein supply for animal feed, and the poultry industry is the largest consumer of such feed. Two kilograms of grain must be fed to poultry to produce 1 kg of weight gain, much less than that required for pork or beef. However, for every gram of protein consumed, chickens yield only 0.33 g of edible protein.

## Economic Factors

Changes in commodity prices for poultry feed have a direct effect on the cost of doing business in the poultry industry. For instance, a significant rise in the price of corn in the United States can put significant economic pressure on large industrial chicken farming operations.

## Worker Health and Safety

Poultry workers experience substantially higher rates of illness and injury than manufacturing workers do on average.

## Muscular Disorders

For the year 2013, there were an estimated 1.59 cases of occupation-related illness per 100 full time U.S. meat and poultry workers, compared to .36 for manufacturing workers overall. Injuries are associated with repetitive movements, awkward postures, and cold temperatures. High rates of carpal tunnel syndrome and other muscular and skeletal disorders are reported. Disinfectant chemicals and infectious bacteria are causes of respiratory illnesses, allergic reactions, diarrhea, and skin infections.

## Respiratory Consequences

Poultry housing has been shown to have adverse effects on the respiratory health of workers, ranging from a cough to chronic bronchitis. Workers are exposed to concentrated airborne particulate matter (PM) and endotoxins (a harmful waste product of bacteria. In a conventional hen house a conveyor belt beneath the cages removes the manure. In a cage-free aviary system the manure coats the ground, resulting in the build-up of dust and bacteria over time. Eggs are often laid on the ground or under cages in the aviary housing, causing workers to come close to the floor and force dust and bacteria into the air, which they then inhale during egg collection.

## Excretory Consequences

Oxfam America reports that huge industrialized poultry operations are under such pressure to maximize profits that workers are denied access to restrooms.

## World Chicken Population

The Food and Agriculture Organization of the United Nations estimated that in 2002 there were nearly sixteen billion chickens in the world, counting a total population of 15,853,900,000. The figures from the *Global Livestock Production and Health Atlas* for 2004 were as follows:

1. China (3,860,000,000)
2. United States (1,970,000,000)
3. Indonesia (1,200,000,000)
4. Brazil (1,100,000,000)
5. Pakistan (691,948,000)
6. India (648,830,000)
7. Mexico (540,000,000)
8. Russia (340,000,000)
9. Japan (286,000,000)
10. Iran (280,000,000)
11. Turkey (250,000,000)
12. Bangladesh (172,630,000)
13. Nigeria (143,500,000)

In 2009 the annual number of chicken raised was estimated at 50 billion, with 6 billion raised in the European Union, over 9 billion raised in the United States and more than 7 billion in China.

In 1950, the average America consumed 20 pounds of chicken per year, but it is predicted that the average consumption will be 89 pounds in 2015. Additionally, in 1980 most chickens were sold whole, and by 2000 almost 90 percent of chickens were sold after being processed into parts. This increase in consumption and processing has led to many of these occupation-related illness.

## Broiler Industry

A broiler chicken in Ecuador

The Broiler industry is the process by which broiler chickens are reared and prepared for meat consumption.

### Broiler Industry Structure

The broiler production process is very much an industrial one. There are several distinct components of the broiler supply chain.

### Primary Breeding Sector

The "primary breeding sector" consists of companies that breed pedigree stock. Pedigree stock ("pure line") is kept on high level biosecure farms. Their eggs are hatched in a special pedigree hatchery and their progeny then goes on to the great grandparent (GGP) and grandparent (GP) generations. These eggs would then go to a special GP hatchery to produce Parent Stock (PS) which passes to the production sector.

In 2006, out of an estimated world population of 18 billion poultry, about 3% are breeding stock. The US supplied about 1/4 of world GP stock.

Worldwide, the primary sector produced 417 million parent stock (PS) per year.

A single pedigree-level hen might have 25000 parent stock bird descendents, which in turn might produce 3 million broilers.

Numerous techniques are used to assess the pedigree stock. For example, birds might be examined with ultrasound or x-rays to study the shape of muscles and bones. The blood oxygen level is measured to determine cardiovascular health. The walking ability of pedigree candidates is observed and scored.

The need for high levels of R&D spending prompted consolidation within the primary breeder industry. By the late 2000s only three sizable breeding groups remained:

- Aviagen (with the Ross, Arbor Acres, Indian River and Peterson brands)
- Cobb-Vantress (with the Cobb, Avian, Sasso and Hybro brands), and
- Groupe Grimaud (with the Hubbard and Grimaud Frere brands).

In the UK, 2 international firms supply about 90% of the parent stock.

Due to the high levels of variation in the chicken genome, the industry has not yet reached biological limits to improved performance.

The full chicken genome was published in Nature, in December 2004. Today, all primary breeding groups are investing heavily in genomics research. This research mostly focuses on understanding the function and effect of genes already present in the breeding population. Research into transgenics — removing genes or artificially moving genes from one individual or species to another — has fewer prospects of gaining favor among consumers.

## Broiler Breeder (Parent Stock) Farms

Broiler breeder farms raise parent stock which produce fertilized eggs. A broiler hatching egg is never sold at stores and is not meant for human consumption. The males and females are separate genetic lines or breeds. The chicks they produce will therefore be hybrids or 'crosses'. Since the birds are bred mainly for efficient meat production, producing eggs can be a challenge. In Canada, the average producer houses 15,000 birds that begin laying hatching eggs at 26 weeks of age. Each bird will lay about 150 hatching eggs for the next 34 to 36 weeks. This cycle is then repeated when the producer puts another flock of 26 week-old birds into his barns to begin the process again. As a general rule, each farmer produces enough broiler hatching eggs to supply chicks for 8 chicken producers. (Other sources indicate a parent hen will lay about 180 eggs in 40 week production period.)

Generally, parent flocks are either owned by integrated broiler companies or hatcheries or are contracted to them on a long-term basis.

Broiler breeder growing is typically a two-stage process. Parent stock purchased from a primary breeder is delivered as day old. Most are first placed with on specialist *rearing houses* or starter farms until approximately 18 weeks of age. The starter farm has the specialized brooding equipment to raise the chicks.

## Rearing House

A typical rearing house (also called a shed or barn) design for Alabama-like climate (100 °F (38 °C) in summer and 20 °F (−7 °C) in winter):

- 40 by 400 feet (12 m × 122 m) size, single storey.
- 11,000 bird capacity (about 1.4 sq ft (0.13 $m^2$) per bird)
- Ceiling is insulated
- Exterior curtain side walls.
- A "minimum ventilation" system is required for the heating period to provide a certain amount of fresh air.
- A separate "tunnel ventilation" system with evaporative pad cooling is desired (minimum wind speed is 400 fpm) for hot weather in the later stage of the bird's growth.
- Air inlets may be automatically adjusted.
- A negative ventilation system helps keep dirt and dust out of egg storage areas.
- The entire house may be heated, or individual "brooders" may be used.
- The floor is flat. There are no "slats" or "pits" for manure. There are no cages, and no nests. "Litter" (shavings or straw) covers the floor. When the chicks are introduced temporary barriers are used to keep them close to the heated areas.
- "Black-out" design to keep out external light, so the day-night cycle can be controlled.
- An automatic timer-controlled lighting system. Dimmers allow light intensity to be adjusted.
- Automatic feeders to distribute feed. Typically this consists of an endless chain in a trough or with individual pans. A silo or bin outside provides storage.
- Automatic drinkers provide water. There are several different designs, with "nipples" or "round" drinkers being popular.
- Feeders and drinkers are height adjusted as the birds grow, and can be raised on chains or wires to allow cleanout of the barn.

Chicks require warm air temperatures, which is reduced as the birds mature:

| Age | Brooder Temperature | Whole-House Heating Temperature |
|---|---|---|
| 0 days | 34–35 °C (93-95 °F) | 31–32 °C (88-90 °F) |
| 14 days | 31–32 °C (88-90 °F) | 24–25 °C (75-77 °F) |

Chicks might be debeaked at 7–10 days age. During rearing, bird weight is carefully monitored, as an over-weight bird will be a poor egg producer. The feed mix will be adjusted to meet nutritional needs at each stage. Feed might be restricted to control body weight, for example with "skip a day" feeding, or feeding 5 days out of 7. A vaccination program is carried out, which ensures the longevity of the parent stock, and the immunity may be passed to the broiler progeny. Males (cockerels) and females (pullets), are usually raised separately.

## Laying House

The birds are then moved to broiler breeder *laying houses* or production barns. The birds are typically placed into crates, and transported by truck to a separate facility. Males and females are raised together at this point. Outwardly the laying house will resemble the rearing house. Inside, about one-half of the floor might consist of raised 'slats.' During the production run, manure will drop through the slats and accumulate in the pit underneath the slats. The birds are not generally caged, especially since the roosters must mate with the hens to fertilize the eggs. Nests are provided for laying hens. Both automatic and manual (example) nesting systems exist. Manual nests are usually stuffed with straw or shavings and eggs are hand-collected. Automatic systems usually have a plastic carpet lining, with a belt for egg collection. Careful layout and attention to bird behavior is required to avoid 'floor eggs'.

Depending on breed, egg production starts at 24–26 weeks of age. Production percentage (daily eggs per hen) climbs rapidly to a peak of 80–85% at 29–32 weeks, and then gradually declines with age. Hatchability tends to peak (at perhaps 90%) somewhat later than production at 34–36 weeks. Overall flock production will decline as mortality reduces the size of the flock.

When the rooster mates with the hen, sperm enter the hen's oviduct and are stored within sperm storage glands. These glands can store more than half a million sperm, and sperm can remain viable for up to 3 weeks. However, a hen will have maximum fertility for only about 3 to 4 days after one mating. Therefore, the male-to-female ratio in a flock must be enough to ensure mating of every hen every 3 days or so. To maintain fertility, younger roosters may be introduced as the flock ages.

Eggs are collected a minimum of twice a day, and usually more frequently. Cracked or dirty eggs are separated, as they are not suitable for hatching. Undersized, oversized or double-yolk eggs are also unsuitable. The eggs might be disinfected by fumigation, are packed in 'flats' or trays, placed in wheeled trolleys, and stored in a cool (15-18C) climate-controlled area. The egg packing room and storage rooms are kept segrated to reduce contamination. The trolleys are delivered by truck to a hatchery perhaps twice a week.

At the end of the production cycle, the birds are called "spent fowl". Disposal of spent fowl may be a problem as consumer demand for them is poor.

## Hatcheries

Hatcheries take the fertilized eggs, incubate them, and produce day old broiler chicks.

Incubation takes about 21 days, and is often a two-step process. Initial incubation is done in machines known as *setters*. A modern setter is the size of a large room, with a central corridor and racks on either side. Eggs are held relatively tightly (large end up) in trays, which are stored in the racks. Inside the setter, temperature and humidity are closely maintained. Blowers or fans circulate air to ensure uniform temperature, and heating or cooling is applied as needed by automated controls. The racks pivot or tilt from side to side, usually on an hourly basis. As an example, one commercial machine can hold up to 124,416 eggs and measures about 4.7 metres wide by 7.2 metres deep. Setters often hold more than one hatch, on a staggered hatch-day basis, and operate continuously. The setter phase lasts about 18 days.

Five-day-old broiler strain Cornish-Rock chicks

On or about day 18, the eggs are removed from the setters and transferred to *hatchers*. These machines are similar to setters, but have larger flat-bottom trays, so the eggs can rest on their sides, and newly hatched chicks can walk. Having a separate machine helps keep hatching debris out of the setter. The environmental conditions in the hatcher are optimized to help the chicks hatch. As a commercial example, a large hatcher has capacity for 15,840 eggs, and measures about 3.3 metres by 1.8 metres.

Some incubators are single-stage (combining setter and hatcher funcations), and entire trolleys of eggs can be rolled in at one time. One advantage of single-stage machines is that they are thoroughly cleaned after each hatch, whereas a setter is rarely shutdown for cleaning. The single-stage environment can be adjusted for eggs from different producing flocks, and for each step of the hatch cycle. The setter environment is often a compromise as different egg batches are in the machine at one time.

On hatch day (day 21), the trays are removed ("pulled") from the hatchers, and then the chicks are removed from the trays. Chicks are inspected, with the sickly ones being disposed of. Chicks may be by vaccinated, sorted by sex, counted, and placed in chick boxes.(Example1) (Example2) Stacks of chick boxes are loaded into trucks for transport, and arrive at the broiler farm on the same day. Specialized climate-controlled trucks are typically used, depending on climate and transport distance.

Chick sexing is optionally done to improve uniformity – since males grow faster, the weights are more uniform if males and females are raised separately. The birds are bred so that males and females have unique feather patterns or color differences. Unlike egg-laying poultry, males are not culled.

Typical hatchability rate in Canada in 2011 was 82.2%. (i.e. 82.2% of eggs set for incubation produced a saleable chick). A UK source estimates 90% hatchability.

## Broiler Farms

Broiler chickens in a farm

The chicks are delivered to the actual broiler *Grow-Out* farms. In the US, houses may be up to 60' x 600' (36000 sq.ft.). One 2006 magazine survey reported a desired 67 foot wide house, with the average 'standard' new house being 45' x 493', with largest being 60' x 504'. One farm complex may have several houses.

In Mississippi, typical farms now have four to six houses with 25,000 birds per house. One full-time worker might manage three houses. On average, a new broiler house is about 500 feet long by 44 feet wide and costs about $200,000 equipped.

When the birds are full-grown, they are caught (perhaps with a Chicken harvester) placed in crates, and transported by truck to a processing plant.

Broiler chickens kept outside near a chicken shop in India

Because of their efficient meat conversion, broiler chickens are also popular in small family farms in rural communities, where a family will raise a small flock of broilers.

## Processing Plants

When the birds are large enough, they are shipped to processing plants for slaughter. When chickens arrive at the processor they go through the following sequence:

- Removed from transport cages
- Hung by the legs on a shackle, mounted on a conveyor chain.
- Stunned using an electrically charged water bath
- Killed by cutting the blood vessels in the neck
- Bled so that most blood has left the carcass
- Scalded to soften the attachment of the feathers
- Plucked to remove the feathers
- Head removed
- Hock cutting to remove the feet
- Rehung in the evisceration room
- Gutted or eviscerated to remove the internal organs
- Washed to remove blood and soiling from the carcass
- Chilled to prevent bacterial spoiling (They go through a chiller which takes approximately 2 hours to go through. The chiller generally holds thousands of

gallons of water kept below 40 degrees Fahrenheit.)

- Drained to allow excess water to drip off the carcass
- Weighing
- Cut selection to divide the carcass into desired portion (breast, drumsticks etc.)
- Packed (for example in plastic bags) to protect carcasses or cuts
- Chilled or frozen for preservation

Further Processing plants carry out operations such as cutting and deboning. Previously the conveyor belts carrying live chickens generally ran at a maximum of 140 chickens per minute, but the maximum speed has been increased to 175 birds/minute. Once the dead birds arrive in the evisceration room (usually dropped down a chute after the feet are removed), they are hung again on shackles much the same way as they were when they were alive.

## Feed Mills

## Integrators

Today, in the U.S. an individual company called an "integrator" performs all or most production aspects. Integrators generally own breeder flocks, hatcheries, feed mills, and processing plants. The integrators provide the chicks, feed, medication, part of the fuel for brooding, and technical advisers to supervise farm production. Integration reduces costs by coordinating each stage of production.

## U.S. Industry History

In the 1920s–1930s, broiler production was initiated in locations such as the Delmarva Peninsula, Georgia, Arkansas, and New England. Mrs. Wilmer Steele of Sussex County, Delaware, is often cited as the pioneer of the commercial broiler industry. In 1923, she raised a flock of 500 chicks intended to be sold for meat. Her business was so profitable that by 1926 she was able to build a 10,000 bird broiler house.

In 1945, A&P organized the first of its "Chicken of Tomorrow" contests. Qualifying trials were conducted in 1946 and 1947 with the national finals held in 1948. Breeders submitted a case of 30 dozen hatching eggs to a hatchery, the eggs were hatched, the offspring raised until they reached market weight and were then slaughtered. Broilers were judged on several factors, including growth rate, feed conversion efficiency, and the amount of meat on breasts and drumsticks. Though held only three times, the contests enabled breeders such as Peterson, Vantress, Cobb, Hubbard, Pilch and Arbor Acres to become established market brands.

During the 1940s – 1960s, feed mills, hatcheries, farms, and processors were all separate entities. Hatcheries were driven to co-ordinate activities to protect their market share and production. Later, feed mills extended credit to farmers to purchase feed to produce the live chickens. Eventually entrepreneurs consolidated feed mill, hatchery and processing operations, resulting in the beginnings of the integrated industry.

Chickens were typically sold "New York dressed," with only the blood and feathers removed. In 1942, an Illinois plant was the first to win government approval of "on-line" evisceration. Evisceration and ice-packing of ready-to-cook whole carcasses became the norm. In 1949, USDA launched a voluntary program of grading. Federal inspection of broilers became mandatory in 1959.

By 1952, "broilers" surpassed farm chickens as the number one source of chicken meat in the United States.

By the mid-1960s, ninety percent of broilers produced came from vertical integrated operations, which combined production, processing and marketing.

In the late 1960s and early 1970s, major companies used television and print media to market chickens under brand names. Today, 95 percent of broilers sold at retail grocery stores carry a brand name.

By the early 1980s, consumers preferred cut-up and further-processed chickens to the traditional whole bird.

Chicken passed pork consumption in 1985. Chicken consumption surpassed beef consumption in 1992.

| | **Historic** | **Modern** |
|---|---|---|
| Number of Hatcheries | 11,405 (1934) | 323 (2001) |
| Incubator Capacity | 276 million eggs (1934) | 862 million eggs (2001) |
| Hatchery Average Incubation Capacity | 24,224 eggs (1934) | 2.7 million eggs (2001) |
| Annual Broiler Production | 366 million broilers (1945) | 8.4 billion broilers (2001) |
| Average live weight | 3.03 pounds (1945) | 5.06 pounds (2001) |
| Live weight price | 36 cents per pound (1948) | 39.3 cents per pound (2001) |
| Feed Conversion Efficiency | 4.70 lbs feed per lb live weight (1925) | 1.91 lbs feed per lb live weight (2011) |
| Mortality | 18% (1925) | 3.8% (2011) |

## Breeders

Cobb claims to be world's oldest poultry breeding company. Founded 1916 when Robert C. Cobb Senior purchased a farm in Littleton, Massachusetts, forming Cobb's Pedigreed Chicks. Purchased by Upjohn in 1974. Sold to Tyson Foods in 1994.

Hubbard was founded by Oliver Hubbard in 1921 in Walpole, New Hampshire. Acquired by Merck in 1974. In 1997 Hubbard was spun off and merged with the ISA-group from France as part of Merial. In 2003, split from ISA, while keeping the broiler lines from ISA and Shaver. Sold by Merial to Groupe Grimaud in 2005.

Arbor Acres was originally a family farm, started by Italian immigrant Frank Saglio who purchased a Glastonbury, Connecticut farm in 1917. He started raising chickens in abandoned piano crates. His third son Henry Saglio took over the poultry while in grade eight. Henry began trying to breed a white bird, because black pinfeathers were difficult to pluck. In 1948, and again in 1951, Arbor Acres White Rocks won in the purebred category of the Chicken of Tomorrow competition. The white feathered Arbor Acres birds were preferred to the higher performing dark feathered Red Cornish crosses. In 1964, Nelson Rockefeller purchased Arbor Acres making it part of International Basic Economy Corporation (IBEC). Joint ventures were formed in Thailand, Taiwan, Indonesia, India, the Philippines, and Japan. In 1980, IBEC merged with Booker McConnell Limited of Great Britain. Booker owned all of "AA" by 1991. At this time Arbor Acres had grown to become the world's largest broiler breeding company, with customers in over 70 countries. AA was divested in 2000, eventually acquired by Avigen.

Shaver started with 2 hens in 1932 by Donald Shaver. Mainly focused on laying hens, Shaver launched a broiler product in 1958. Cargill purchased part of Shaver in 1964, which helped give Shaver a toehold in the US market. In the early 1970s the market share in the US was around 8-10%. Cargill bought all of Shaver in 1985. Shaver was acquired by ISA in 1988, and then made part of Merial. The layer business kept the Shaver name, and was sold as Natexis Industrie in 2003, and then to Hendrix in 2005.

## Industry Statistics

### World

Worldwide, from 1985 to 2005, the broiler industry grew by 158%. Major increases were experienced by:

- China +591%
- Brazil +482%
- US +147%
- Thailand +141%
- EU-25 +73%.

In 2005 world production was 71,851,000 tonnes. Major producers were:

- United States 15,869,000 tonnes

- China 10,196,000 tonnes
- EU-25 8,894,000 tonnes
- Brazil 8,668,000 tonnes

In 2005, world exports of chicken meat $8.3 billion (CAD). Largest exporters were Brazil ($4 billion), the United States ($2.6 billion) and the EU-25 ($0.82 billion). The largest importers of chicken meat were: Japan ($1 billion), Russia ($943.3 million), Germany ($800.6 million) and Hong Kong ($598.8 million).

## United States

In 2010, approximately 36.9 billion pounds (16,737,558 tonnes) of broilers were sold, for a retail value of $45 billion, based on retail weight sold multiplied by the retail composite price. In 2010, the US exported 6.8 billion pounds, valued at $3.1 billion, about 18% of production.

In 2009, the US produced 8.6 billion birds. The top 3 states were Georgia, Arkansas and Alabama, each producing over 1 billion birds. Farm receipts were about $22 billion.

There are fewer than 50 highly specialized, vertically integrated agribusiness firms that dominate the industry. The top 10 integrators produce about 60% of U.S. broilers.

In 2001, there were 323 chicken hatcheries, with an incubator capacity of 862 million eggs. The average capacity per hatchery was 2.7 million eggs.

In 2010 the largest producers were Tyson Foods (161 million ready to cook pounds) and Pilgrim's Pride (126.5 million pounds). The next largest producer, Perdue Farms, is less than half the size of Pilgrims Pride.

## Canada

Canada has a supply management system where marketing boards govern the broiler and broiler hatching egg industries. For broilers, prices are negotiated at the provincial level. In each province, the minimum price per kg that processors will pay to producers is set periodically through negotiations between processors and the provincial marketing board. From 1992 to 2003, negotiated prices in Ontario are generally used as a benchmark when conducting price negotiations in other provinces. In Ontario, Chicken Farmers of Ontario (CFO) has price-negotiating authority. It negotiates the base price paid by primary processors for live chicken with primary processors. Since 2003, the live chicken price is determined by a "live price formula" established by the Agriculture, Food, and Rural Affairs Appeals Tribunal that includes the price of chicks, feed and producer margin.

Broiler hatching egg production consists of 270 producers generating about $188.3 million in 2005. Canada produced about 675 million hatching eggs, and imported about 121 million.

There were 66 hatcheries in Canada, of which 20 were mixed, producing both broiler and layer chicks. The main companies involved in broiler hatching eggs and chicks are:

- Maple Leaf Foods Incorporated,
- Lilydale Hatchery,
- Maple Lodge,
- Couvoir Boire & Frères Inc
- Western Hatchery Limited.

The average price per chick was about $0.35. Canada imported about 13 million broiler chicks.

There were 2786 regulated chicken producers, generating farm cash receipts of $1.6 billion in 2005. Compared to other livestock sectors (i.e. beef, dairy, and pork), the poultry and egg industry was the healthiest with regards to total income for the average operator.

In 2005, total chicken slaughters were 973.9 million kilograms. Of this, 35.2 million kg were mature (non-broiler)slaughters, meaning about 96% of chicken consumption was broilers. By revenue, chicken processing is about 1/4 of the meat packing business. The top 8 processors account 66% of the market.

In 2005 there were 175 primary poultry processing plants. The five largest firms are, in order:

1. la Coopérative fédérée de Québec (three plants in Québec),
2. Lilydale Poultry Co-operative (one plant in British Columbia, three in Alberta and one in Saskatchewan),
3. Maple Leaf Poultry (two plants in Ontario, one in Alberta and one in Nova Scotia),
4. Exceldor (two plants in Québec) and
5. Maple Lodge Farms (one plant in Ontario).

There are 376 plants that do further processing, involving canning, boning and cutting.

| Chicken consumption by market sector | Consumption (000,000 kg) |
|---|---|
| Retail (Grocery Stores, Butcher Shops | 625 |
| Fast Food | 231 |
| Full Serve Restaurants | 97 |
| Hotels, Institutions | 55 |
| Total | 1008 |

The Canadian Food Inspection Agency (CFIA), is a branch of Health Canada whose role is to enforce the food safety, to ensure animal health, to set standards and carry out enforcement and inspection. Activities range from the inspection of federally registered meat processing facilities to border inspections for foreign pests and diseases, to the enforcement of practices related to fraudulent labelling. The CFIA also verifies the humane transportation of animals, conducts food investigations and recalls, and performs laboratory testing and environmental feeds.

The Canadian On-Farm Food Safety Program (COFFSP) is directed by producers. It is a science-based, credible program consistent with the Hazard Analysis Critical Control Points (HACCP, "hass ap") standards, managed by the Canadian Federation of Agriculture.

## References


- Appleby, M.C.; J.A. Mench; B.O. Hughes (2004). Poultry Behaviour and Welfare. Wallingford and Cambridge MA: CABI Publishing. ISBN 0-85199-667-1.
- Lester R. Brown (2003). "Chapter 8. Raising Land Productivity: Raising protein efficiency". Plan B: Rescuing a Planet Under Stress and a Civilization in Trouble. NY: W.W. Norton & Co. ISBN 0-393-05859-X.
- Brulliard, Karin (10 June 2016). "Egg producers pledge to stop grinding newborn male chickens to death". The Washington Post. Retrieved 12 June 2016.
- Ogle, Maureen. "Riots, Rage, and Resistance: A Brief History of How Antibiotics Arrived on the Farm". Scientific American. Sep 3, 2013. Retrieved 28 October 2016.
- Grow, et al. "Farmaceuticals: The drugs fed to farm animals and the risks posed to humans", Reuters, Filed Sept. 15, 2014, 1 p.m. GMT. Retrieved on 28 October 2016.
- U.S. FDA. [ http://www.fda.gov/ForConsumers/ConsumerUpdates/ucm378100.htm "Phasing out certain antibiotics use in farm animals"], Dec 11, 2013. Updated Feb 25, 2015. Retrieved on 28 October 2016.
- Sneeringer, et al. "Economics of Antibiotic Use in U.S. Livestock production", USDA, November 2015. Retrieved on 28 October 2016.
- Pelton, Tom. "New Book Explains How Perdue's 'Chickenizing" Changed the World", 'WYPR', Sep 8, 2016. Retrieved on 28 October 2016.

- Stephanie Strom (July 31, 2015). "Perdue Sharply Cuts Antibiotic Use in Chickens and Jabs at Its Rivals". The New York Times. Retrieved August 12, 2015.
- Adler, Jerry; Lawler, Andrew (June 2012). "How the Chicken Conquered the World". Smithsonian. Retrieved April 19, 2015.
- Yashroy, Rakesh. "Poultry production under Salmonella stress: Infection mechanisms". Research Gate. Retrieved November 18, 2014.

# 3

# Essential Concepts in Poultry Science

The essential concepts of poultry science are capon, feed conversion ratio, broodiness, plumping, organic egg production, poultry show etc. A capon is a rooster which has been castrated whereas plumping is the process by which some companies inject raw chicken meat with saltwater or with chicken stock. The following section provides a glimpse of the essential concepts in poultry science.

## Poultry Litter

In agriculture, poultry litter or broiler litter is a mixture of poultry excreta, spilled feed, feathers, and material used as bedding in poultry operations. This term is also used to refer to unused bedding materials. Poultry litter is used in confinement buildings used for raising broilers, turkeys and other birds. Common bedding materials include wood shavings, sawdust, peanut hulls, shredded sugar cane, straw, and other dry, absorbent, low-cost organic materials. Sand is also occasionally used as bedding. The bedding materials help absorb moisture, limiting the production of ammonia and harmful pathogens. The materials used for bedding can also have a significant impact on carcass quality and bird performance.

There are specific practices that must be followed to properly maintain the litter and maximize the health and productivity of the flocks raised on it. Many factors must be considered in successful litter management including time of the year, depth of the litter, floor space per bird, feeding practices, disease, the kind of floor, ventilation, watering devices, litter amendments, and even the potential fertilizer value of the litter after it is removed from the house. Most poultry are grown on dirt floors with some type of bedding material. Concrete floors and some specialized raised flooring are used at some facilities. In many areas of the United States, shavings from pine or other soft woods have historically been the bedding of choice for poultry production. Regionally, other materials have been the bedding material of choice due to regional cost and availability, such as rice hulls in the lower Mississippi River poultry production areas of Arkansas and Mississippi.

### Bedding Materials

Growers consider a number of factors when determining which material to use as bedding in their facilities, with cost and availability being a major consideration. Bedding materials generally needs to be very absorbent, and must have a reasonable drying time. Many paper products, for instance, absorb moisture well but do not dry out appropriately. The material should also have a useful purpose once it has been used as a bedding material. Without a useful purpose for the used litter, poultry growers would need to dispose of unmanageable quantities of old litter. Large accumulations of litter stored unused for long periods of time are not ecologically acceptable even on a small scale, and would be non-sustainable from an industrial perspective.

Poultry bedding materials also have to be reasonably available. Some materials may meet industry goals once under the birds but if it is difficult to obtain, it will not find favor as a poultry litter. Finally, if a material is not cost competitive with current materials utilized, it will also not be used as a litter material. However, if the new material has increased value once removed from the poultry house compared to current litters or if the current litter material itself becomes difficult to obtain or the quality is decreases, poultry growers may decide to use the new litter material.

Bedding material must not be toxic to poultry or to poultry growers. It should not be excessively favorable for the growth of the litter beetle, a major pest. The effect on other livestock, pets, wildlife, and even plants must also be considered. Poultry can consume as much as 4% of their diet as litter, therefore any bedding material must not contain contaminants, such as pesticides or metals. Consumption by the birds due to litter eating or other bird behavior could affect production and potentially cause the meat or rendered products to become unusable. Pine shavings has been the bedding of choice because of performance, availability, and cost.

## Management Practices

### Moisture

The heating and ventilation systems in a poultry house must be continuously monitored to keep the moisture content of the litter controlled so that the litter remains friable (easily broken up or crumbly). If the litter becomes too wet and the litter is allowed to become "sealed", then the birds will be living on a damp, slippery and sticky surface. This sealed litter is what is referred to as being "caked." In this condition, the litter is simply saturated with water and is unable to dry out. A severe problem with litter moisture will result if large areas of the house floor surface are caked. The more common issue, however, is having localized areas of caking near leaky watering cups, nipples, troughs or roofs. Watery droppings caused by nutrition and/or infectious agents can also be a cause of excessive moisture in poultry litter.

If litter is not kept at an acceptable moisture level, very high bacterial loads and unsanitary growing conditions may result producing odors (including ammonia), insect problems (particularly flies), soiled feathers, footpad lesions and breast bruises or blisters. This can affect the health and mortality rate of the flock, and could result in quality issues when birds reared under such poor conditions reach the processing plant. In a well-managed broiler house, litter moisture normally averages between 25 and 35 percent. Litter that is managed correctly with the moisture content kept within the acceptable range can be reused if no disease or other production problems occur. On the other hand, caked litter must be removed between flocks and replaced with new litter.

## Litter Re-utilization

Some broiler producers are simply removing cake and excess litter after house washing and then placing broilers on old litter for an extended number of flocks. Their expectation is that total clean out is not needed unless there some disease or other bio-security issues. However, producers doing this should be aware that total disinfection under these conditions is probably not possible.

Re-utilization of at least some fraction of used litter as a supplement for fresh wood shavings bedding in broiler houses has been found to not significantly increase pathogens and indicator microorganisms in litter compared to using fresh wood shavings. No consistent significant differences have been found regarding flock performance when comparing houses using fresh litter with houses re-utilizing litter.

A major issue with re-utilization of previously used litter is the generation of ammonia. Ammonia is produced by microbial breakdown of fecal material in the litter. It is well documented in the literature that higher moisture levels result in higher ammonia production. The caked portion of the litter is very high in moisture and nitrogen and should be removed from the house to reduce ammonia generation and provide optimal air quality for chicks during the brooding period. Add litter treatments to reduced ammonia generation. Controlling ammonia with a litter treatment can save money on energy costs by reducing the amount of air exchange required to maintain adequate air quality.

## Litter Amendments

High ammonia levels in poultry houses can result in poor bird performance and health and a loss of profits to the grower and integrator. When broilers and turkeys are raised on litter, amendments can be used to reduce ammonia levels in the houses and improve productivity. Uric acid and organic nitrogen (N) in the bird excreta and spilled feed are converted to ammonium ($NH_4^+$) by the microbes in the litter. Ammonium, a plant-available N form, can bind to litter and also dissolve in water. Depending on the moisture content, temperature, and acidity of the litter, a portion of the ammonium will be converted into ammonia ($NH_3$). Ammonia production is favored by high temperature and high pH (i.e., alkaline conditions). Ammonia is a pungent gas that irritates the eyes and respiratory system and can reduce resistance to infection in poultry. At high-enough concentrations, ammonia will reduce feed efficiency and growth while increasing mortality and carcass condemnations. The result is economic loss to the grower and integrator. Because chicks are more susceptible to the negative effects of ammonia, placing broods in houses with high levels of builtup litter is particularly harmful. Also, the high temperatures required during brooding increase ammonia levels, and moist litter (due to leaky drinkers or high water tables) and insufficient winter ventilation contribute to high ammonia levels as well. In these situations, some growers rely mainly on ventilation to reduce ammonia in the houses. However, ammonia loss from the litter reduces its fertilizer value, and venting ammonia into the environment can cause health and environmental problems.

There are several types of litter amendments available to manage ammonia the most common being acidifiers, and various microbial and enzymatic treatments.

## Acidifiers

This type of amendment creates acidic conditions (pH less than 7) in the litter, resulting in more of the ammoniacal nitrogen being temporarily retained as ammonium rather than ammonia. Ammonium is a highly reactive ion that bonds with sulfates, nitrates and phosphates to form ammonium salts that improve the nutrient value of litter when land applied as fertilizer. The acidity also creates unfavorable environment for urolytic bacteria reducing the production of enzymes that contribute to ammonia formation, resulting in reduced ammonia production. Urolytic bacteria have a pH optimum of approximately 8.3 and litter amendments lower the litter surface pH to below 4.0 for a short period of time, usually between 3–5 days depending on the litter amendment. There are several different types of acidifiers, such as alum, acidified liquid alum, sodium bisulfate, ferric sulfate, and sulfuric acid, that have been used by the poultry industry. These products vary in effectiveness as the pH is raised by the activity of the poultry within a couple of weeks. The combination of using litter amendments and poultry house ventilation provides a healthy and comfortable rearing environment. Controlling volatilized ammonia in poultry and livestock rearing environments is critical to maintaining a high level of animal health, well-being and efficient live performance. All of these products are regulated by DOT and HAZMET and physical properties for the different products rang from mild irritant to corrosive. All litter acidifiers are corrosive but when applied according to manufacturers instructions are safe and effective.

## Other Amendments

There have been a number of other substances used for ammonia control. A study in Finland found that peat, which is high in humic acid, when used as poultry litter it was quite effective in controlling ammonia. A number of products have also appeared on the market using de-nitrifying or nitrogen-fixing bacteria.

## Windrowing

One common practice is windrow composting. This is a deep stacking of litter, usually by plowing the litter into long rows the length of the poultry house. This is an incomplete composting process, and can eliminate harmful pathogens such as E. coli and Salmonella providing that the internal stack temperature reaches 140 to 160 °F. Re-spreading the stacked litter and allowing it to dry would be expected to decrease ammonia and extend litter life.

## Disposal and Re-use

Broilers have on average a 47-day growout period, during which the typical broiler chicken will generate about two pounds of litter, if you add the manure and bedding materials. Actual manure generation will be lower because it is only a fractional component of litter. This translates to an average of about 0.7 ounce per day per bird, varying considerably over the life of the bird. This means that a single broiler house, which can contain well over 20,000 birds can generate over 40,000 lbs of litter per flock.

Historically, applications for used poultry litter have included, and still include, use as feed for cattle in the commercial beef industry, land application as a fertilizer for crops or pastures, or occasionally as potting material for the greenhouse and plant container industries. Recently there has been an upsurge in the use of poultry litter as a bio-fuel source for electrical cogeneration and gasification.

### Use as Fertilizer

Poultry litter's traditional use is as fertilizer. As with other manures, the fertilizing value of poultry litter is excellent, but it is less concentrated than chemical fertilizers, giving it a relatively low value per ton. This makes it uneconomical to ship long distances, and it tends to lose its nitrogen value fairly quickly. Extracting its value requires that it be used on nearby farms. This limits its resale value in regions where there are more poultry farms than suitable nearby farmland.

### Use as Cattle Feed

Traditionally used as fertilizer, it is now also used as a livestock feed as a cost-saving measure compared with other feedstock materials, particularly for beef animals.

The use of poultry litter as food for beef cattle is legal in the United States. Prior to 1967, the use of poultry litter as cattle feed was unregulated but that year the FDA issued a policy statement that poultry litter offered in interstate commerce as animal feed was adulterated, effectively banning the practice. In 1980, FDA reversed this policy and passed regulation of litter to the states. In December 2003, in response to the detection of bovine spongiform encephalopathy (mad cow disease) in a cow in the state of Washington, the FDA announced plans to put in place a poultry litter ban. Because poultry litter can contain recycled cattle proteins as either spilled feed or feed that has passed through the avian gut, the FDA was concerned that feeding litter would be a pathway for spreading mad cow disease. In 2004, FDA decided to take a more comprehensive approach to BSE that would remove the most infectious proteins from all animal feeds. The FDA decided at this point that a litter ban was unnecessary in part based on comments by the North American Rendering Industry. In 2005, the FDA published a proposed rule that did not include a litter ban and in 2008 the final rule did not include the ban either.

### Use as Fuel

There are currently several electrical generating plants in the UK, and recently in the US, that are utilizing poultry and turkey litter as their primary fuel. The first three (and the world's first three of these plants) were developed by Fibrowatt Ltd in the UK, founded by Simon Fraser, who was appointed an OBE for his contribution to renewable energy. These are: Thetford (38.5 MWe), Eye (12.7 MWe) and Glanford (13.5 MWe - now switched to burning meat and bonemeal). The fourth, Westfield (9.8 MWe), was developed by Energy Power Resources, which now owns all four. Simon Fraser's son and partner, Rupert Fraser, went on to develop the first US plant through his company Fibrowatt LLC., at Benson, MN (55 MWe).

On a smaller scale, poultry litter is used in Ireland as a biomass energy source. This system uses the poultry litter as a fuel to heat the broiler houses for the next batch of poultry being grown thus removing the need for LPG gas or other fossil fuels.

Some companies are also developing gasification technologies to utilize poultry litter as a fuel for electrical and heating applications, along with producing valuable by-products including activated carbons and fertilizers.

## Capon

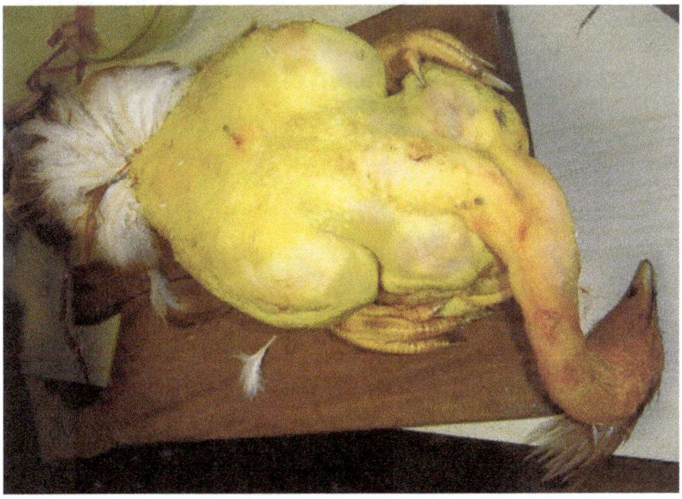

A plucked capon with its head, feet and tail feathers still attached.

A capon (from Latin *caponem*) is a cockerel or rooster that has been castrated to improve the quality of its flesh for food and, in some countries like Spain, fattened by forced feeding.

In the United Kingdom, birds sold as capons are chemically or physically castrated cocks.

Live capons in Hainan, China. Note the small head, comb and wattle

## History

The origins of caponised chickens are contested. Such culinary practices existed far-east within ancient China as well as Europe from Greek and Roman empires.

One of the earliest records of caponisation occurred during the time of the Roman Republic. The Lex Faunia of 162 BC forbade fattening hens to conserve grain rations. To get around this the Romans instead castrated roosters, which resulted in a doubling of size. It was also practiced later throughout medieval times with gastronomic texts describing capons as preferred poultry since the ordinary fowl of the farmyard was regarded as peasant fare and "popular malice crediting monks with a weakness for capons."

The practice of caponisation found its way in modern-day America through the east of Philadelphia where the industry is predominant. Today the nation of France is internationally renowned for maintaining a strong caponisation tradition with widespread and established industries throughout the country.

William Shakespeare mentioned capon in the famous "All the world's a stage" monologue from his play *As You Like It* (written c.1600). He similarly describes capon as a food of the wealthy. The monologue describes human life as consisting of seven stages, and the fifth stage is a middle-aged man who has reached the point where he has acquired wisdom and wealth. The monologue describes the fifth stage as: "The Justice, In fair round belly, with a good capon lin'd".

## Effects of Caponisation

Caponisation is the process of turning a cockerel into a capon. Caponization can be done by surgically removing the bird's testes, or may also be accomplished through the use of estrogen implants. With either method, the sex hormones normally present are no longer effective. Caponization must be done before the rooster matures, so that it develops without the influence of sex hormones.

Capons, due to the lack of male sex hormones, are not as aggressive as normal roosters. This makes capons easier to handle and allows capons to be kept together with other capons since their reduced aggressiveness prevents them from fighting.

The lack of sex hormones results in meat that is less gamey in taste. Capon meat is also more moist, tender and flavorful than that of a cockerel or a hen, which is due not only to the hormonal differences during the capon's development but also because capons are not as active as roosters, which makes their meat more tender and fatty.

Capons develop a smaller head, comb and wattle than those of a normal rooster.

Capons are fairly rare in industrial meat production. Chickens raised for meat are bred and raised so that they mature very quickly. Industrial chickens can be sent to market in as little as five weeks. Capons produced under these conditions will taste very similar to conventional meat, making their production unnecessary.

### Specialized Production

Capon is produced in France, in many regions, notably Gers and Jura regions. It is a specialty of Bresse (Chapon de Bresse), where it has its own appellation to differentiate it from other capons from other regions. In Bresse, it is exclusively produced from the Bresse blue foot breed (patte bleue), and fed a certain diet which makes it even more tender than from other regions and breeds, making it the world's best, most sought after and expensive chicken.

It is also produced in many regions of northern Italy, such as Piedmont, Friuli-Venezia Giulia, Emilia-Romagna and Marche. In the early 20th century, capon was commonly eaten for the main family feast on Christmas in Tuscany and northern Italian regions, which was a rare treat for peasant or working class families.

## Feed Conversion Ratio

In animal husbandry, feed conversion ratio (FCR) or feed conversion rate is a measure of an animal's efficiency in converting feed mass into increases of the desired output. For dairy cows, for example, the output is milk, whereas animals raised for meat – such as beef cows, pigs, chickens, and fish – the output is the mass gained by the animal, the final mass of the animal, or the mass of the dressed output. Specifically FCR is the mass of the input divided by the output. In some sectors feed efficiency, which is the output divided by the input (the inverse of FCR), is used.

## Background

Feed conversion ratio (FCR) is the ratio of inputs to outputs; it is the inverse of "feed efficiency" which is the ratio of outputs to inputs. FCR is widely used in hog and poultry production, while FE is used more commonly with cattle. Being a ratio the FCR is dimensionless, that is, it is not affected by the units of measurement used to determine the FCR.

FCR a function of the animal's genetics and age, the quality and ingredients of the feed, and the conditions in which the animal is kept, and storage and use of the feed by the farmworkers.

As a rule of thumb, the daily FCR is low for young animals (when relative growth is large) and increases for older animals (when relative growth tends to level out). However FCR is a poor basis to use for selecting animals to improve genetics, as that results in larger animals that costs more to feed; instead Residual Feed Intake (RFI) is used which is independent of size. RFI uses for output the difference between actual intake and predicted intake based on an animal's body weight, weight gain, and composition.

The outputs portion may be calculated based on weight gained, on the whole animal at sale, or on the dressed product; with milk it may be normalized for fat and protein content.

As for the inputs portion, although FCR is commonly calculated using feed dry mass, it is sometimes calculated on an as-fed wet mass basis, (or in the case of grains and oilseeds, sometimes on a wet mass basis at standard moisture content), with feed moisture resulting in higher ratios.

### Conversion Ratios for Livestock

Animals that have a low FCR are considered efficient users of feed. However, comparisons of FCR among different species may be of little significance unless the feeds involved are of similar quality and suitability.

### Beef Cattle

As of 2013 in the US, a FCR calculated on live weight gain of live-weight gain of 4.5–7.5 was in the normal range with an FCR above 6 being typical. As of 2013 FCRs had not changed much compared to other fields in the prior 30 years, especially compared to poultry which had improved feed efficiency by about 250% since the late 1800s.

### Dairy Cattle

The dairy industry traditionally didn't use FCR but in response to increasing concentration in the dairy industry and other livestock operations, the EPA updated its regulations in 2003 controlling manure and other waste releases produced by livestock operators. In response the USDA began issuing guidance to dairy farmers about how to control inputs to better minimize manure output and to minimize harmful contents, as well as optimizing milk output.

In the US, the price of milk is based on the protein and fat content, so the FCR is often calculated to take that into account. Using an FCR calculated just on the weight of protein and fat, as of 2011 an FCR of 13 was poor, and an FCR of 8 was very good.

Another method for dealing with pricing based on protein and fat, is using energy-corrected milk (ECM), which adds a factor to normalize assuming certain amounts of fat and protein in a final milk product; that formula is (0.327 X milk pounds) + (12.95 X fat pounds) + (7.2 X protein pounds).

In the dairy industry, feed efficiency (ECM/intake) is often used instead of FCR (intake/ECM); an FE less than 1.3 is considered problematic.

FE based simply on the weight of milk is also used; an FE between 1.30 and 1.70 is normal.

## Pigs

As of 2011, pigs used commercially in the UK and Europe had an FCR, calculated using weight gain, of about 1 as piglets and ending about 3 at time of slaughter. As of 2012 in Australia and using dressed weight for the output, a FCR calculated using weight of dressed meant of 4.5 was fair, 4.0 was considered "good", and 3.8, "very good". In the US as of 2012, commercial pigs had FCR calculated using weight gain, of 3.46 for while they weighed between 240 and 250 pounds, 3.65 between 250 and 260 pounds, 3.87 between 260 and 270 lbs, and 4.09 between 280 and 270 lbs.

Because FCR calculated on the basis of weight gained gets worse after pigs mature, as it takes more and more feed to drive growth, countries that have a culture of slaughtering pigs at very high weights, like Japan and Korea, have poor FCR ratios.

## Sheep

Some data for sheep illustrate variations in FCR. A FCR (kg feed dry matter intake per kg live mass gain) for lambs is often in the range of about 4 to 5 on high-concentrate rations, 5 to 6 on some forages of good quality, and more than 6 on feeds of lesser quality. On a diet of straw, which has a low metabolizable energy concentration, FCR of lambs may be as high as 40. Other things being equal, FCR tends to be higher for older lambs (e.g. 8 months) than younger lambs (e.g. 4 months).

## Poultry

As of 2011 in the US, broiler chickens has an FCR of 1.6 based on body weight gain, and mature in 39 days. At around the same time the FCR based on weight gain for broilers in Brazil was 1.8.

For hens used in egg production in the US, as of 2011 the FCR was about 2, with each hen laying about 330 eggs per year.

From the early 1960s to 2011 in the US broiler growth rates doubled and their FCRs halved, mostly due to improvements in genetics and rapid dissemination of the improved chickens. The improvement in genetics for growing meat created challenges for farmers who breed the chickens that are raised by the broiler industry, as the genetics that cause fast growth decreased reproductive abilities.

### Crickets

Crickets have a low feed conversion ratio of only 1.7.

### Fish

The FIFO ratio (or Fish In - Fish Out ratio) is the feed conversion ratio applied to aquaculture, where the first number is the mass of harvested fish used to feed farmed fish, and the second number is the mass of the resulting farmed fish. A ratio of 3:1 would mean that for every kg of fish farmed, 3 kg fish (generally of another, cheaper to produce or capture fish species) is used to farm it.

As of 2015 farm raised Atlantic salmon had a commodified feed supply with four main suppliers, and an FCR of around 1 Tilapia is about 1.5, and as of 2013 farmed catfish had a FCR of about 1.

### Rabbits

In India, rabbits raised for meat had an FCR of 2.5 to 3.0 on high grain diet and 3.5 to 4.0 on natural forage diet, without animal-feed grain.

## Broodiness

Broodiness is the action or behavioral tendency to sit on a clutch of eggs to incubate them, often requiring the non-expression of many other behaviors including feeding and drinking. Being broody has been defined as "Being in a state of readiness to brood eggs that is characterized by cessation of laying and by marked changes in behavior and physiology". Broody birds often pluck feathers from their chest and abdomen, using them to cover the eggs. As a consequence of this, they develop one or several patches of bare skin on the ventral surface. These reddish, well-vascularized areas of skin are usually called brood patches, and improve heat transfer to the eggs. Broodiness is usually associated with female birds, although males of some bird species become broody and some non-avian animals also show broodiness.

A brooding White tern (*Gygis alba*).

## In Wild Birds

In wild birds, egg incubation is a normal and essential phase in the process of reproduction, and in many families of birds, e.g. pigeons, the eggs are incubated by both male and female parents.

## Broodiness in Males

In all the species of phalaropes the males become broody rather than the female. The females leave the nest after finishing laying to let the males incubate the eggs and take care of the young. Male emus (*Dromaius novaehollandiae*) become broody after their mates start laying, and begin to incubate the eggs before the laying period is complete.

## Non-broodiness

A small number of atypical birds such as *Passeriformes* of the genus *Molothrus* do not become broody but lay their eggs in the nests of other species for incubation, known as brood parasitism. The Australian Brushturkey *(Alectura lathami)* also does not become broody, rather, it covers the eggs with a large mound of vegetable matter which decomposes keeping the eggs warm until hatching. The Crab-plover, (*Dromas ardeola*), which live on the coasts and islands of the Indian Ocean, let their eggs incubate primarily by the heat of the sun and will leave their nests unattended, occasionally for days at a time.

## In Domestic Poultry

Broody hens can be recognized by their behaviour. They sit firmly over the eggs, and when people approach or try to remove the eggs, threaten the person by erecting their feathers, emitting a characteristic sound like *clo-clo-clo* and will peck aggressively. When broody, hens often temporarily cease eating or reduce their feed consumption.

Letting eggs accumulate in a relatively dark place near the floor often stimulates hens to become broody. Placing artificial eggs into nests also stimulates broodiness. Keeping hens in dark places with warm temperatures and in view of vocalising orphan chicks can induce broodiness, even in breeds that normally do not go broody.

A brooding domestic hen.

Some environmental conditions stimulate broodiness. In heavy breeds of chickens, warm weather tends to bring about broodiness. Removing eggs each day, out of the sight of the hens, helps avoid broodiness not only in domestic poultry but also in some wild species in captivity. This continued egg laying means more eggs are laid than would occur under natural conditions. Poultry farming in battery cages also helps to avoid broodiness.

## In Commercial Egg-laying

An egg incubator.

Because hens stop laying when they become broody, commercial poultry breeders perceive broodiness not as a normal physiological process, but as a drawback to egg and poultry meat production. With domestication, it has become more profitable to incubate eggs artificially, while keeping hens in full egg production. To help achieve this, there has been intense artificial selection for non-broodiness in commercial egg laying chickens Gallus gallus domesticus and parent stock of poultry. As a result of this artificial selection, broodiness has been greatly reduced to very low levels in present breeds of commercial fowl, both among egg-laying and meat-producing breeds.

## Physiological Basis

Broodiness is due to the secretion of prolactine hormone by the front lobe of the hypophysis. Prolactine injection in hens provokes egg laying to stop within a few days, vitellum reabsorbtion, ovary regression (hens only have a left ovary) and finally broodiness. However, attempts to stop broodiness by the administration of several hormones have failed because this state, once evoked, requires time to revert.

Prolactine injections inhibit the production of gonadotropin hormone, a hormone that stimulates ovarian follicles which is produced in the frontal lobe of hypophysis.

Castrated males can go broody with baby chicks, showing that broodiness is not limited to females, however, castrated males do not incubate eggs.

Contrary to common opinion, the temperature of broody hens barely differs from that of laying hens. Broody hens pluck feathers from their chest, using them to cover the eggs. As a consequence of this, they develop one or several patches of bare skin on the ventral surface. These reddish, well-vascularized areas of skin are usually called brood patches. which improve heat transfer to the eggs.

## Genetic Basis

Broodiness is more common in some chicken breeds than others, indicating that it is an inherited characteristic. Breeds such as Cochin, Cornish and Silkie exhibit a tendency to broodiness, including brooding eggs from other species such as quails, pheasants, turkeys and geese. In some breeds such as the White Leghorn, broodiness is extremely rare.

Some studies on crosses of chicken breeds point to the hypothesis of complementary genes acting on broodiness. Other results point to the hypothesis of sex-linked genes, or, inheritance through the maternal chromosome. Although these studies have been made on different breeds of chickens, their results are not contradictory. There is common agreement that artificial selection for egg production succeeded in reducing the incidence of broody hens in chicken populations.

## Chicken Breeds that Commonly Exhibit Broodiness

- Cochin
- Manx Rumpy
- Plymouth Rock
- Silkie
- Saipan Jungle Fowl
- Orpington
- Kraienköppe
- Belgian Bearded d'Anvers
- Icelandic chicken
- Java (chicken)
- Philippine Native Chicken
- Belgian Bearded d'Uccle
- Iowa Blue
- Nankin
- Delaware
- Booted Bantam
- New Hampshire
- Pekin
- Dutch Bantam
- Indian Game (Cornish)

## Chicken Breeds that Rarely Exhibit Broodiness

- Leghorn
- Polish
- Norwegian Jærhøne
- Brabanter

- Orloff
- Sultan
- Scots Grey
- Sussex
- Sebright
- Menorca
- Wyandotte
- Egyptian Fayoumi

## Broodiness in Non-avian Animals

There is some evidence that non-avian dinosaurs also practiced brooding. A specimen of the Mongolian oviraptorid *Citipati osmolskae* was discovered in a chicken-like brooding position in 1993, which may indicate that they had begun using an insulating layer of feathers to keep the eggs warm.

Several deinonychosaur and oviraptorosaur specimens have also been found preserved on top of their nests, likely brooding in a bird-like manner.

Lungless salamanders in the family *Plethodontidae* lay a small number of eggs in a cluster among damp leaf litter. The female salamander often broods the eggs and in the genus *Ensatinas*, she has been observed to coil around them and press her throat area against them, effectively massaging them with a mucous secretion. The black mountain salamander mother broods her eggs, guarding them from predation as the larvae feed on the yolks of their eggs. They eventually break their way out of the egg capsules and disperse. Some species of Gymnophiona (caecilians, with long, cylindrical, limbless bodies) brood their eggs.

A brooding female python.

Most pythons coil around their egg-clutches and remain with them until they hatch. A female python will not leave the eggs, except to occasionally bask in the sun or drink water. She will even "shiver" to generate heat to incubate the eggs.

Some cichlid fish lay their eggs in the open, on rocks, leaves, or logs. Male and female parents usually engage in differing brooding roles. Most commonly, the male patrols the pair's territory and repels intruders, while females fan water over the eggs, removing the infertile and leading the fry while foraging. However, both sexes are able to perform the full range of parenting behaviours.

## Mouthbrooding

Mouthbrooding, also known as oral incubation, refers to the care given by some groups of animals to fertilized eggs or their offspring by holding them in the mouth of the parent for extended periods of time. Although it has been observed in a variety of animals, most mouthbrooders are fish. The parent performing this behavior invariably feeds less often and afterwards will be underweight, requiring a period of feeding and restoring the depleted energy reserves.

## Others

Marsupial frogs are so-called because they possess a dorsal brood pouch. In some species the eggs are fertilized on the female's lower back, and are inserted in her pouch with the aid of the male's toes. The eggs remain in contact with the female's vascular tissue, which provides them oxygen.

Some animals have a common name that includes the word 'brood' or its derivatives, although it is arguable whether the animals show 'broodiness' per se. For example, the female gastric-brooding frog (*Rheobatrachus sp.*) from Australia, now probably extinct, swallows her fertilized eggs, which then develop inside her stomach. She ceases to feed and stops secreting stomach acid and the tadpoles rely on the yolks of the eggs for nourishment. After six or seven weeks the mother opens her mouth wide and regurgitates the tadpoles which hop away from her mouth. The Brooding sea anemone (*Epiactis prolifera*) is a colonial hermaphrodite that fertilizes and incubates its eggs internally. The motile larvae, after swimming out of the mouth, migrate down to the disk and become fixed there until they become little anemones, ready to move and feed independently.

In Darwin's Frog (*Rhinoderma darwinii*), the female lays about 30 eggs and then the male guards them for about two weeks, until they hatch. The male then takes all the survivors and carries around the developing young in his vocal pouch. When the tiny tadpoles have developed they hop out and swim away. In this animal, the parents hold the hatched young rather than eggs in their mouths, so is arguably not showing 'broodiness'.

# Plumping

Plumping, also referred to as "enhancing" or "injecting," is a term that describes the process by which some poultry companies inject raw chicken meat with saltwater, chicken stock, seaweed extract or some combination thereof. The practice is most commonly used for fresh chicken and is also used in frozen poultry products, although other meats may also be plumped.

Poultry producers have injected chicken (and other meats) with saltwater solutions since the 1970s, claiming it makes for tastier, juicier meat. According to Kenneth McMillin, Professor of Meat Science at the Louisiana State University Agricultural Center in Baton Rouge, processors use multiple-needle injectors or vacuum-tumblers, that force the sodium solution into the muscle. Binding agents in the solution prevent the added salt and water from leaching out of the meat during transport, in grocery stores and during cooking.

## Controversy

### Cost to Consumers

Plumped chicken commonly contains 15% of its total weight in saltwater, but in some cases can contain as much as 30%. Since the price of chicken is based on weight, opponents of the practice estimate that shoppers could be paying up to an additional $1.70 per package for added saltwater, with the total annual cost to U.S. families estimated to be $2 billion in added weight charges.

### Health Effects

A serving of plumped chicken can contain between 200 mg and 500 mg of sodium per serving, which is more than 25% of the USDA's recommended daily sodium intake. Non-plumped chicken generally contains 45 to 70 mg per serving. In January 2010, the American Heart Association released new guidelines calling for all Americans to reduce their sodium intake to 1,500 mg (equivalent to 3.8 g of salt) from 2,300 mg. Previously, 1,500 mg was the recommended limit for higher-risk individuals only., In a current study, research has shown that reducing salt intake by three grams a day would decrease new cases of heart disease by one-third each year. This would reduce heart disease- related deaths by an estimated 100,000 annually, and save up to an estimated $24 billion in annual health costs, according to a study published by the New England Journal of Medicine. Dr. Bibbins-Domingo of UCSF, who led the study for the New England Journal of Medicine says, "Reducing salt intake could be as beneficial as quitting smoking, weight loss, and using cholesterol medication."

Supporters of plumping cite that chicken with as much as 330 mg or less of sodium per serving is eligible for the American Heart Association's seal of approval. However, the

National Academy of Sciences Institute of Medicine recently conducted hearings exploring health risks associated with high salt intake, including testimony that addressed the practice of plumping. Health experts have weighed in on the issue. In a June 22, 2009 *Los Angeles Times* article, Marion Nestle, Professor of Nutrition, Food Studies and Public Health at New York University, noted that "the practice of saltwater plumping [adds] unnecessary salt to people's diets, [and] it also increases the water weight of chicken."

## Labeling

Poultry producers, consumer advocacy groups and government officials are debating how plumped chicken products should be labeled. Under current USDA regulations, plumped chicken may still carry an "All-Natural" or "100% Natural" label. Plumped chicken is commonly labeled as natural and "enhanced with up to 15% chicken broth." Critics of the practice argue that 70% of consumers think chicken labeled "natural" should not contain saltwater. Proponents of plumping say that their packaging already states that the chicken has been injected and that the enhanced chicken is sold to meet consumer demand.

In South Africa, controversy over individually quick-frozen chicken pieces being injected with up to 50% brine content, has resulted in government regulation requiring IQF chicken to be labeled with brine content. IQF chicken is a cheap staple food and manufacturers were accused of misleading the poor.

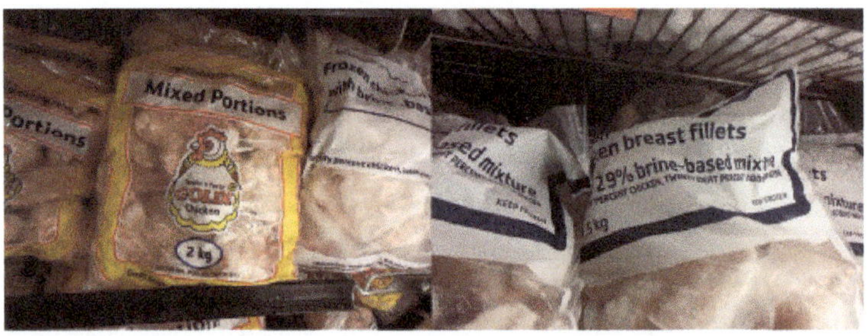


IQF chicken pieces in South Africa (from two different manufacturers), labeled with brine content. The labels consistently assert that the product is only 70% chicken.

## Media Coverage

A number of U.S. poultry producers have addressed the practice in advertising campaigns. In March 2009, West Coast poultry producer Foster Farms and their advertising agency Goodby, Silverstein & Partners launched a campaign centered on "Say No To Plumping." TV commercials in English and Spanish featured the long-running Foster "Imposters" discussing the effects of plumping. The "Say No To Plumping" campaign won the 2010 Gold Effie Award in advertising. Sanderson Farms previously ran an advertising campaign on TV and outdoor billboards drawing consumer attention to saltwater injected chicken.

Plumping gained attention from the Wall Street Journal in early May 2009. The issue has also been covered in the Los Angeles Times, the Associated Press, the Baton Rouge Advocate and the Deseret News in Salt Lake City, Utah.

## Organic Egg Production

Sussex Hen outdoors.

Organic egg production is the production of eggs through organic means. In this process, the poultry are fed organic feed. According to the United States Department of Agriculture, organic means that the laying hens must have access to the outdoors and cannot be raised in cages. Organic egg producers cannot use antibiotics except during an infectious outbreak. Only natural molting can occur within the flock; forced molting is not allowed. Organic certification also requires maintenance of basic animal welfare standards.

### Differences Between "Free Range" and "Organic"

Significant differences cover feed, medication, and animal welfare. Organic hens are fed organic feed; it is prohibited to feed animal byproducts or GMO crops – which is not disallowed in free range environments; no antibiotics allowed except in emergencies (in free range, it is up to the farmer, but the same levels of antibiotics as conventional farming is allowed); required animal welfare standards in organic farms are higher, which can improve the quality of both the eggs and the meat.

### Organic Feed

Organic feed is grown by certified organic farmers. To become a certified organic farmer, the crop must be free of genetically modified organisms (GMOs). The crop must be free of GMOs and synthetic fertilization for three years before it can be certified for organic usage. If the crop is contaminated by cross-fertilization, the crop is rendered useless for organic grading. Finally, there can be no animal by-products fed to the poultry.

### Living Conditions

At an organic farm (in Bruthen, Victoria) chickens sometimes end up laying eggs somewhere other than the owner expected

In the United States, "organic" egg production means that the flock may not live in cages and must have access to outdoor areas.

### Antibiotics

Organic egg producers cannot feed low-level antibiotics to the poultry. Antibiotics are only allowed during an outbreak of infection or disease.

### Molting

Some farms induce molting in their flocks to affect egg production. In organic egg farms, the birds are allowed to go into a natural molt but are not induced.

### Animal Welfare

Similar to all other forms of egg production in the United States, organic production is also regulated by animal welfare audit system. Mistreatment of the chickens could potentially lead a farmer to losing their organic certification. Thus some of the arguments from animal activist organizations that egg production is cruel and inhumane do not necessarily apply to organic raised hens. On the other hand, male chicks who are born on organic or free-range egg farms are still discarded, by the use of lethal gas, because they do not produce eggs. The American Organic Standards still allow beak trimming as a means to lessen injury to the birds.

## Incubator (Egg)

An incubator is a device simulating avian incubation by keeping eggs warm and in the correct humidity, and if needed to turn them, to hatch them. Reginald Carl A. Sanchez invented this incubator for the eggs even if there is no hen to hatch it.

## Styles

Modern incubators are electrically heated with a thermostat. Incubators can be used in a farmhouse, such as a large chicken raising facilities, or they can be found in a common classroom for students to observe the egg inside and when it hatches. Some industrial incubators are large enough to hold up to 124.416 eggs , while some other styles can only hold a few eggs.

The different styles of incubators include the following:

- Setter Incubator
- hatcher incubator
- combination incubator

## Purposes

The incubator is an apparatus that is used for environmental conditions, such as temperature and humidity that needs to be controlled. It is often used for growing bacterial cultures, hatching eggs artificially, or providing suitable conditions for a chemical or biological reaction. The incubator is recorded to hatch not only bird eggs, but it also is used to hatch reptile eggs. It allows the fetus inside of the egg to grow without the mother needing to be present to provide the warmth. Chicken eggs are recorded to hatch after about 21 days, but other species of birds can either take a longer or shorter amount of time. An incubator is supposed to be able to set the perfect environment and condition for an egg to incubate because it regulates the factors such as temperature, humidity, and turning the eggs when necessary. This is so that the egg incubated properly because it plays the role of the hen in its natural state. The incubator also allows the egg to incubate while eliminating the external threats that could possibly harm the eggs. The modern hatchery manager's goal is to produce large numbers of uniform, robust day-old chicks. Robustness is a health criterion, originating in the embryonic life stage of the chicken – and correlating directly with the performance and resistance of individual chicks under differing farm conditions. It is possible to incubate different species of birds at the same time within the same incubator.It was also used to raise birds

## Incubation Methods

In industrial incubation, there are two common used methods of incubation. In single-stage incubation, the incubator contains only eggs of the same embryonic age. The advantage of single-stage incubation is that climate conditions can be adjusted according to the needs of the growing embryo. In multi-stage incubation the setter contains eggs of different embryonic ages. Usually 6 or 2 age groups. Consequently, climate conditions cannot exactly be adjusted according to the needs of all the growing embryos and a compromise has to be sought to best suit the age groups presented in the setter. In a multi-stage incubation procedure, the heat produced by the older embryos is utilized to heat up the warmth demanding younger embryos For Single-stage and multi-stage incubation, not necessarily different incubators are needed. Some machines are able to run both methods based on their programming.

## Common Names

The common names of the incubator in other terms include the following:

- breeding machines
- hatching machines
- hatchers
- setters
- artificial incubation equipment
- egg breeding equipment
- egg hatching equipment
- brutgerate
- brutmaschinen
- broeimasjiene
- broeikaste
- eier broeiers
- aanbroeiers
- voorbroeiers
- uitbroeiers
- kunsmoeders
- sikilatör

## History

A magazine published in Britain during WWII described an incubator as "a wooden box, hot water, and a curtain". One of the first recorded methods of incubating included using the heat of rotted manure to warm the eggs. The Egyptians had a better method of incubating that used a cylindrical building that had a fire at the bottom of the building. The eggs that were incubating were placed on an inverted cone that was partially covered in ash. The eggs were placed in a woven basket that sat on top of the ashes. The building also had a roof that allowed smoke to escape, but it kept the rain out. In 400 B.C. Mechanical incubating was not invented until the year of 1749 by Reamur in Paris, France. Lyman Byce created a coal lamp incubator in 1879. The first commercial machine was made by Hearson in the year of 1881.

## Poultry Show

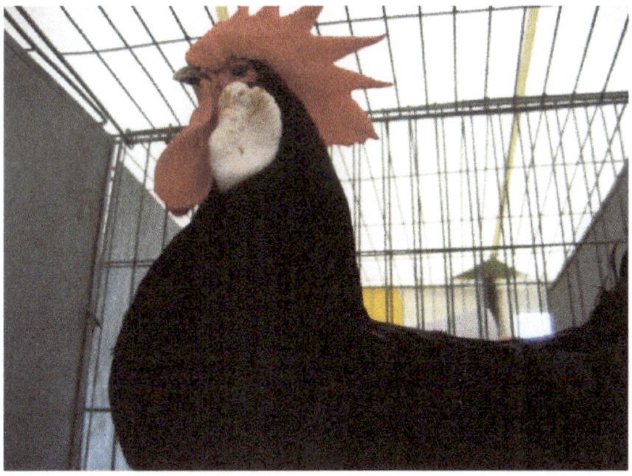

A Minorca breed rooster at a Scottish poultry show.

A poultry show is a specific subset of a livestock show that involves the exhibition and competition of exhibition poultry, which may include chickens, domestic ducks, domestic geese, domestic guineafowl and domestic turkey. Domestic pigeon are also exhibited but not universally considered poultry. As well as being independent events, they are also sometimes held in conjunction with agricultural shows.

### Preparation

Significant effort is put into exhibiting poultry. Birds are trained for the cages used at shows, and washed and manicured in preparation for an event.

## History

The first poultry show in the United Kingdom was in 1845 in London. The exhibition of poultry was promoted as an alternative to cock fighting in the United Kingdom following the banning of such activities in 1849. The first poultry standard in the world was produced in 1865, the British Poultry Standard. The height of the poultry showing in the United Kingdom was during the late 19th and early 20th century, when The Crystal Palace Poultry Show were held at The Crystal Palace, the largest attracting 10,533 filled cages, the actual number of birds unknown as many were pairs. Charles Darwin regularly attended and showed pigeons at these from 1857, and Queen Victoria exhibited her Cochins in 1846 at the Royal Dublin Show. The Crystal Palace Great Poultry Shows continued to run with great success until the Crystal Palace burnt down.

The first poultry show in North America was held on 15-16 November, 1849, in Quincy Market, Boston, surpassing expectations to receive over 1400 entries however it proved a debacle as no poultry standard existed and judging was cancelled as there was no way to decide what was a purebred. A poultry association was organised afterwards and another show attempted the next year, which attracted 12,000 birds. Overcrowding led to judging being cancelled again.

In 1854 P. T. Barnum staged a *National Poultry Show* at his Barnum's American Museum, which the guidelines of are unknown. It wasn't until 1874 that the American Standard of Perfection were adopted by the recently formed American Poultry Association, creating the first poultry standard in North America.

## Governance

Poultry shows are generally organised by a poultry club. Many of these are accredited by a national body, such as the Poultry Club of Great Britain while others, such as Australian poultry clubs are only managed on a state level. Birds exhibited at poultry shows are judged to the local standard, a compiled manual that describe the ideal form of a poultry breed.

## References


- Maguelonne Toussaint-Samat (trans.: Anthea Bell) (2009). "The History of Poultry". The History of Food (revised ed.). John Wiley & Sons. pp. 305–15. ISBN 978-1444-30514-2.
- Carole M. Counihan (2004). Around the Tuscan Table: Food, Family, and Gender in Twentieth Century Florence. Routledge. p. 61. ISBN 9781135939632. Retrieved December 15, 2014.
- Linda Civitello (2007). Cuisine and Culture: A History of Food and People (2 ed.). John Wiley & Sons. p. 265. ISBN 9780471741725.
- DJ Cottle and WS Pitchford. Production Efficiency. Chapter 18 in Beef Cattle Production and Trade, Ed Lewis Kahn. Csiro Publishing, 2014 ISBN 9780643109896.

- Collavo, A. et al. House cricket small-scale farming. In Ecological implications of minilivestock: potential of insects, rodents, frogs and snails, Ed M.G. Paoletti. CRC Press 2005. ISBN 9781578083398.
- Stebbins, Robert C.; Cohen, Nathan W. (1995). A Natural History of Amphibians. Princeton University Press. p. 196. ISBN 0-691-03281-5.
- Dorit, R. L.; Walker, W. F.; Barnes, R. D. (1991). Zoology. Saunders College Publishing. pp. 853–854. ISBN 0-03-030504-7.
- Rude, Emelyn (2016). Tastes Like Chicken: A History of America's Favorite Bird. Wiley. ISBN 978-1-68177-163-2. Retrieved 25 August 2016.
- Dohner, Janet Vorwald (2001). The encyclopedia of historic and endangered livestock and poultry breeds. New Haven: Yale University Press. p. 411. ISBN 9780300138139. Retrieved 28 March 2015.
- "The Journal of Agriculture and Horticulture". Council of Agriculture of the Province of Quebec. 15 January 1907. Retrieved 28 March 2015.
- Steup, H.H (1926). Fitting and Exhibiting Standard - Bred Poultry : Volume 127 of Kansas Agri. Exp. Station Circular (PDF). Kansas Agricultural Experiment Station. Retrieved 28 March 2015.
- Lacey, Patricia (2010). All Cooped Up: The History of the American Bantam Association A 131-Year Evolution. pp. 17–22. ISBN 9781450060837. Retrieved 28 March 2015.

# 4

# Methods and Techniques of Poultry Farming

The methods and techniques discussed within this chapter are free range, yarding, battery cage, furnished cages, debeaking, forced molting etc. Animals that can roam around free for some part of the day are denoted by the word free range where as yarding is the custom of fencing animals in order to keep them safe. Yarding is a method of intensive grazing. This text discusses the methods and techniques of poultry farming in a critical manner providing key analysis to the subject manner.

## Free range

Commercial free range hens

Free range denotes a method of farming husbandry where the animals, for at least part of the day, can roam freely outdoors, rather than being confined in an enclosure for 24 hours each day. On many farms, the outdoors ranging area is fenced, thereby technically making this an enclosure, however, free range systems usually offer the opportunity for extensive locomotion and sunlight prevented by indoor housing systems. *Free range* may apply to meat, eggs or dairy farming.

The term is used in two senses that do not overlap completely: as a farmer-centric description of husbandry methods, and as a consumer-centric description of them. There is a diet where the practitioner only eats meat from free-range sources called ethical omnivorism, which is a type of semivegetarian.

In ranching, free-range livestock are permitted to roam without being fenced in, as opposed to fenced-in pastures. In many of the agriculture-based economies, free-range livestock are quite common.

## History

If one allows "free range" to include "herding", free range was a typical husbandry method at least until the development of barbed wire and chicken wire. The generally poor understanding of nutrition and diseases before the twentieth century made it difficult to raise many livestock species without giving them access to a varied diet, and the labor of keeping livestock in confinement and carrying all their feed to them was prohibitive except for high-profit animals such as dairy cattle.

A small flock of mixed free-range chickens being fed outdoors.

In the case of poultry, free range was the dominant system until the discovery of vitamins A and D in the 1920s, which allowed confinement to be practised successfully on a commercial scale. Before that, green feed and sunshine (for the vitamin D) were necessary to provide the necessary vitamin content. Some large commercial breeding flocks were reared on pasture into the 1950s. Nutritional science resulted in the increased use of confinement for other livestock species in much the same way.

## United States

In the United States, USDA free range regulations currently apply only to poultry and indicate that the animal has been allowed access to the outside. The USDA regulations do not specify the quality or size of the outside range nor the duration of time an animal must have access to the outside.

The term "free range" is mainly used as a marketing term rather than a husbandry term, meaning something on the order of, "low stocking density," "pasture-raised," "grass-fed," "old-fashioned," "humanely raised," etc.

There have been proposals to regulate by the USDA the labeling of products as free range within the United States. As of now what constitutes raising an animal free range is entirely decided by the producer of that product

## Free-range Poultry

In poultry-keeping, "free range" is widely confused with yarding, which means keeping poultry in fenced yards. Yarding, as well as floorless portable chicken pens ("chicken tractors") may have some of the benefits of free-range livestock but, in reality, the methods have little in common with the free-range method.

Free range meat chickens seek shade on a U.S. farm

A behavioral definition of free range is perhaps the most useful: "chickens kept with a fence that restricts their movements very little." This has practical implications. For example, according to Jull, "The most effective measure of preventing cannibalism seems to be to give the birds good grass range." De-beaking was invented to prevent cannibalism for birds not on free range, and the need for de-beaking can be seen as a litmus test for whether the chickens' environment is sufficiently "free-range-like."

The U.S. Department of Agriculture Food Safety and Inspection Service (FSIS) requires that chickens raised for their meat have access to the outside in order to receive the free-range certification. There is no requirement for access to pasture, and there may be access to only dirt or gravel . Free-range chicken eggs, however, have no legal definition in the United States. Likewise, free-range egg producers have no common standard on what the term means.

The broadness of "free range" in the U.S. has caused some people to look for alternative terms. "Pastured poultry" is a term promoted by farmer/author Joel Salatin for broiler chickens raised on grass pasture for all of their lives except for the initial brooding period. The Pastured Poultry concept is promoted by the American Pastured Poultry Producers' Association (APPPA), an organization of farmers raising their poultry using Salatin's principles.

## Free-range Livestock

Traditional American usage equates "free range" with "unfenced," and with the implication that there was no herdsman keeping them together or managing them in any way. Legally, a free-range jurisdiction allowed livestock (perhaps only of a few named species) to run free, and the owner was not liable for any damage they caused. In such jurisdictions, people who wished to avoid damage by livestock had to fence them out; in others, the owners had to fence them in.

The USDA has no specific definition for "free-range" beef, pork, and other non-poultry products. All USDA definitions of "free-range" refer specifically to poultry.

In a December 2002 Federal Register notice and request for comments (67 Fed. Reg. 79552), USDA's Agricultural Marketing Service proposed "minimum requirements for livestock and meat industry production/marketing claims". Many industry claim categories are included in the notice, including breed claims, antibiotic claims, and grain fed claims. "Free Range, Free Roaming, or Pasture Raised" would be defined as "livestock that have had continuous and unconfined access to pasture throughout their life cycle" with an exception for swine ("continuous access to pasture for at least 80% of their production cycle"). In a May 2006 Federal Register notice (71 Fed. Reg. 27662), the agency presented a summary and its responses to comments received in the 2002 notice, but only for the category "grass (forage) fed" which the agency stated was to be a category separate from "free range." Comments received for other categories, including "free range," are to be published in future Federal Register editions.

## European Union

A free range chicken flock in the USA

The European Union regulates marketing standards for egg farming which specifies the following (cumulative) minimum conditions for the free-range method:

- hens have continuous daytime access to open-air runs, except in the case of temporary restrictions imposed by veterinary authorities,
- the open-air runs to which hens have access is mainly covered with vegetation and not used for other purposes except for orchards, woodland and livestock grazing if the latter is authorised by the competent authorities,
- the open-air runs must at least satisfy the conditions specified in Article 4(1)(3)

(b)(ii) of Directive 1999/74/EC whereby the maximum stocking density is not greater than 2500 hens per hectare of ground available to the hens or one hen per $4m^2$ at all times and the runs are not extending beyond a radius of 150 m from the nearest pophole of the building; an extension of up to 350 m from the nearest pophole of the building is permissible provided that a sufficient number of shelters and drinking troughs within the meaning of that provision are evenly distributed throughout the whole open-air run with at least four shelters per hectare.

Free range geese in Germany

Otherwise, egg farming in EU is classified into 4 categories: Organic (ecological), Free Range, Barn, and Cages.) The mandatory labelling on the egg shells attributes a number (which is the first digit on the label) to each of these categories: 0 for Organic, 1 for Free Range, 2 for Barn and 3 for Cages.

There are EU regulations about what free-range means for laying hens and broilers (meat chickens) as indicated above. However, there are no EU regulations for free-range pork, so pigs could be indoors for some of their lives. In order to be classified as free-range, animals must have access to the outdoors for at least part of their lives.

## United Kingdom

Free range pigs in England

Pigs Free-range pregnant sows are kept in groups and are often provided with straw for bedding, rooting and chewing. Around 40% of UK sows are kept free-range outdoors and farrow in huts on their range.

Egg laying hens Cage-free egg production includes barn, free-range and organic systems. In the UK, free-range systems are the most popular of the non-cage alternatives, accounting for around 28% of all eggs, compared to 4% in barns and 6% organic. In free-range systems, hens are housed to a similar standard as the barn or aviary.

Free-range rearing of pullets Free range rearing of pullets for egg-laying is now being pioneered in the UK by various poultry rearing farms. In these systems, the pullets are allowed outside from as young as 4 weeks of age, rather than the conventional systems where the pullets are reared in barns and allowed out at 16 weeks of age

Meat chickens Free-range broilers are reared for meat and are allowed access to an outdoor range for at least 8 hours each day. Free-range broiler systems use slower-growing breeds of chicken to improve welfare, meaning they reach slaughter weight at 16 weeks of age rather than 5–6 weeks of age in standard rearing systems.

Turkeys Free-range turkeys have continuous access to an outdoor range during the daytime. The range should be largely covered in vegetation and allow more space. Access to fresh air and daylight means better eye and respiratory health. The turkeys are able to exercise and exhibit natural behaviour resulting in stronger, healthier legs. Free-range systems often use slower-growing breeds of turkey.

## Australia

Australian standards in relation to free-range production are largely espoused in third-party certification trade marks due to the absence of any significant legally binding legislation. A number of certification bodies are utilised by rearers to identify their products with a particular level of animal welfare standards. In events where producers do not choose to use a certified trade mark and merely state that their product is 'free range', the producer is bound by consumer expectations and perceptions of what constitutes free range. Producers are generally thought to be bound to Model Codes of Practice of Animal Welfare published by the CSIRO, and in some states this forms part of legislation.

## Egg Laying Hens

In Australia, three farming methods for the production of eggs are utilised. In 2011, traditional cage (or battery) eggs accounted for 42% of value, barn-laid eggs account for 10% of value, free-range eggs accounted for 44% of value, and organic eggs accounted for 4% of value. Increased demand for free range eggs due to customer concerns over animal welfare has led to a number of different standards developing in relation to three core welfare measures - indoor stocking density, outdoor stocking density, and beak trimming. The Model Code of Practice recommends practices for free range farming with the following standards:

- Maximum stocking densities indoors of 30 kg/m2, equivalent to about 14-15 birds/m2.
- Maximum outdoor stocking density of 1500 birds/ha, although this can be increased with rotation onto fresh pasture
- Access to the outdoor range for a minimum of 8 hours per day, except in adverse weather conditions
- 2 metres worth of popholes per 1,000 birds for access to the range
- Beak trimming is permitted, and to be undertaken by an accredited operator

The above standards are not always met, and on some occasions producers may want more ethical standards. As such, certified trade marks play a significant role in the determining of what constitutes free range. The key certifications used for layer hens in Australia include the following...

Egg Corp Assured is the weakest standard, set by the industry peak group and largely based on the Model Code of Practice. Egg Corp Assured differs in that it interprets the outdoor stocking density figure as largely irrelevant to welfare. Egg Corp Assured has been known to certify farms running up to 44,000 birds per hectare outdoors, far in excess of recommendations. Like the Model Code of practice, beak trimming is allowed and indoor densities run up to 15 birds per m2.

RSPCA Approved Farming is a standard that can be applied to both barn-laid and free-range egg producers. Farms using this certification must have an indoor density of 9 birds/m2 indoors on slats, or 7 birds/m2 indoors in a deep-litter system. The standards dictate a maximum outdoor density of 1500 per ha without rotation, or 2500 birds per ha with rotation, and beak trimming is allowed.

Free Range Egg & Poultry Australia (FREPA) standards provides a sliding scale for indoor density, with 10 birds/m2 allowed only in enclosures housing less than 1,000 birds, and 6 birds/m2 the maximum for barns with over 4,000 birds. Nothing is said in the standards about outdoor density, thus it is assumed that farmers must meet the standards under the Model Code. Beak trimming is allowed under this certification.

Humane Choice True Free Range standards are some of the most sound as far as animal welfare is concerned. Beak trimming or any other mutilations is not permitted, perches must be provided, and maximum flock numbers cannot be greater than 2,500 per barn. The outdoor stocking density is 1500 birds per ha, and the indoor density is 5 birds per m2.

Australian Certified Organic Standards include criteria on feed content and the use of pesticides in addition to animal welfare requirements. The indoor density is a maximum of 8 birds/m2, although most operators under this standard list their density as 5 birds per m2. The outdoor density is 1000 birds per ha, and beak trimming is not permitted.

### Chicken Meat

In Australia, free range and organic chicken accounts for about 16.6% of value in the poultry market. This percentage is expected to grow to up to 25% in the next 5 years. No meat birds are raised in cages in Australia. There are three main certification trademarks in this market.

Free Range Egg & Poultry Australia (FREPA) standards are those in which most supermarket brands of free range chicken meat are accredited under. These standards require indoor stocking densities of up to 30 kg per m2 indoors (about 15 birds per m2), and beak trimming is not permitted. Outdoor stocking density is not stated, but it is understood that the outdoor range must be at a minimum 1.5 times the floor area of inside the barn.

RSPCA Approved Farming standards for free range require an indoor stocking density of about 17 birds per m2, and outdoor densities of up to 17 birds per m2. No beak trimming is allowed under this system.

Australian Certified Organic standards dictate a maximum indoor stocking density of up to 12 birds per m2 indoors, and 2500 birds per hectare outdoors. These standards require perches, and prevent large, conventional broiler sheds.

## Yarding

During the daytime, the doors are left open for these chickens to choose whether to be in the yard or coop. This small poultry farm is in Hainan, China.

In poultry keeping, yarding is the practice of providing the poultry with a fenced yard in addition to a poultry house. Movable yarding is a form of managed intensive grazing.

Yarding is often confused with free range. The distinction is that free-range poultry are either totally unfenced, or the fence is so distant that it has little influence on their freedom of movement.

## Historical Practice

Before the discovery of vitamins A and D in the 1920s, green feed and sunshine were essential to the health of poultry. Vitamin D was synthesized from sunlight on the skin (as with humans), while Vitamin A was obtained through green forage plants such as grass. Yards small enough to be fenced economically were soon stripped of palatable green forage and become barren. This is followed by a build-up of manure, parasites, and other pathogens.

Free range husbandry was the most common method in these early days. Most farms had only a small free-range barnyard flock. Larger flocks were kept in small houses build on skids, which were dragged periodically to a fresh piece of ground. This method is similar to the modern practice of pastured poultry.

Experts of the day estimated the sustainable level to be about fifty hens per acre (80 $m^2$ per hen), with one hundred hens per acre (40 $m^2$ per hen) as an absolute upper limit if special care was taken. These levels are sustainable in the sense that the turf can make use of the nutrients in the manure left behind by the chickens, and in the sense that, at this stocking density, the chickens will not completely destroy the turf through scratching.

At the Oregon Station on clay soil it was found that the day droppings from 200 laying hens on an acre [20 $m^2$ per hen] in four years made the soil too rich for the successful growth of cereal crops where cropping the ground was done every other year. The night droppings were put on other land. If the soil contains too much manure for the crops it is safe to assume that it is not in the best condition for poultry. Sooner or later it is bound to show not only in a failure of grain crops but in failure of poultry crops. For a permanent system under average conditions of soil and climate the following points are suggested for consideration.

- Maximum number of fowls per acre: 100 laying hens [40 $m^2$ per hen].
- Disposing of the night droppings on other land.
- Dividing the ground into at least two divisions or yards, and growing a crop on each yard at least every other year. In sections where crops may be grown every year the number of fowls may be increased.
- Growing crops that will use up the maximum amount of manure.
- Keeping the ground vacant [of chickens] at least six months in the year.

- Thorough underdrainage, where necessary, to carry off surplus water.

.... It is not assumed that as many as 500 hens may not be profitably kept on an acre [8 $m^2$ per hen] for a few years under favorable conditions. It has been done, but it is a different matter when it is planned to make a permanent business of it.

Because fifty hens per acre represents 800 square feet (74 $m^2$) per hen (80 $m^2$ per hen), while the density inside the house at the time was normally four square feet per hen (0.4 $m^2$ per hen), this required that the yard be 200 times wider than the house, assuming a yard on one side of the house. That is, a house 20 feet (6 m) wide required a yard 4,000 feet (1,220 m) wide to provide the necessary area. This would normally be provided as two yards, one on either side of the house, each 2,000 feet (610 m) wide. In reality, such yards are expensive to fence, and the chickens spend most of their time on the portion closest to the house, so sustainability was never achieved in practice except with portable houses, which were moved periodically to fresh ground. Yarded operations were operated with unsustainably small yards that were quickly denuded and which received excessive levels of manure.

The use of multiple yards, frequent plowing, and liberal use of lime would allow higher stocking levels to be used, since plowing and liming would allow much of the nitrogen to escape from the soil.

The following is typical advice for the successful use of yards in the Thirties and Forties:

All poultrymen should realize that there are no known substitutes for sunshine and young green grass in keeping poultry in the best possible state of health and in promoting growth and maintaining egg production. Where sunshine and green grass cannot be provided, as in the case of birds kept in strict confinement, the best possible substitutes must be provided. In the case of most farm and many commercial flocks, however, the growing stock is reared on range, and the adult birds are given yards or allowed to roam at will.

If the staggering losses among growing chicks and laying birds that occur annually are to be reduced materially, better methods of flock management must be employed. The losses from mortality are due largely to internal parasites and diseases of one kind or another. Bare ground over which the chickens have run for some time, mud puddles, and stagnant water are the chief sources of the spread of diseases, most of which are filth borne....

The mortality that usually occurs in growing and adult stock may be materially reduced by providing the birds with an alternate yarding system. Probably the best arrangement is to provide each colony brooder and each laying house with three yards (3 m) which the birds would be allowed to use every 3 or 4 weeks. By alternating the birds in the yards every 3 or 4 weeks each yard is kept reasonably sanitary, especially if the soil in the immediate vicinity of the house is cultivated and treated with lime, and young green grass is available for the birds throughout the season. The importance of clean range for both birds and adult stock cannot be emphasized too strongly... For adult stock a good grass sward can be maintained on fertile soil, allowing about 200 birds to the acre [20 $m^2$ per hen].

Nutritional advances increasingly turned yarding into a liability, and it fell out of favor. Free range continued to be used, especially for breeding flocks and for pullets before they reached laying age, because of the lower rate of disease and greater overall health of grass-reared chickens. Breeding flocks (which lay eggs destined for incubation) are always given a better diet than flocks laying table eggs, since a diet that will produce table eggs cheaply will not provide eggs that hatch well. For some time after confined laying flocks produced table eggs satisfactorily, breeding flocks benefited from free range

In Britain, Geoffrey Sykes developed a new yarding system in the Fifties. This used a small yard covered with a thick layer of straw, with more straw added frequently. He also recommended that shade and a windbreak be provided by a solid fence around the yard, or by other means, such as rows of haybales. Once a year, the old straw was removed by a front-end loader or similar machinery. This method eliminated mud and pathogens. It was later forgotten because the industry moved to high-density confinement before the method was widely established.

### Recent Practice

Today, commercial poultry producers generally call yarding free range on their labels. This conflation of two very different techniques has led to confusion. The vast majority of "free-range" operations are really yarded.

Pastured poultry, as promoted by the APPPA, the American Pastured Poultry Producers Association, and author/farmer Joel Salatin, takes a different approach, attempting to achieve the benefits of free range while using penning or yarding. The key element of Pastured poultry is the use of portable housing and the optional use of portable electric fencing. By moving the house and yard frequently, perhaps daily, all the disadvantages of permanent yards are eliminated.

## Battery Cage

Battery cages are a housing system used for various animal production methods, but primarily for egg-laying hens. The name arises from the arrangement of rows and columns of identical cages connected together, sharing common divider walls, as in the cells of a battery. Although the term is usually applied to poultry farming, similar cage systems are used for other animals. Battery cages have generated controversy between advocates for animal rights, and industrial producers.

It was estimated that over 60% of the world's eggs were produced in industrial systems, mostly using battery cages. In the US, over 90% of the 300 million egg-laying chickens are housed in battery cages. In the UK, statistics from the Department for the Environment, Food and Rural Affairs (Defra) indicate that 50% of eggs produced in the UK throughout 2010 were from cages (45% from free-range, 5% from barns). However, introduction of the European Union Council Directive 1999/74/EC which banned conventional battery cages in the EU from January 2012 for welfare reasons, means the number of eggs from battery cages in the EU states is rapidly decreasing. The EU ban was proposed when international scientists independently observed signs of extreme abnormal behaviour (including cannibalism) in caged hens.

## Examples

Battery cages also used for mink, rabbit, chinchilla and foxes in fur farming, and most recently for the Asian palm civet for kopi luwak production of coffee.

Battery cages for sun bears reared for their bile

Battery cages for mink reared for their fur

Battery cages for silver foxes reared for their fur

Battery cages for civets reared for kopi luwak (coffee) production

## History

Before the cage was invented, most hens were free-range.

An early reference to battery cages appears in Milton Arndt's 1931 book, *Battery Brooding*, where he reports that his cage flock was healthier and had higher egg production than his conventional flock. At this early date, battery cages already had the sloped floor that allowed eggs to roll to the front of the cage, where they were easily collected by the farmer and out of the hens' reach. Arndt also mentions the use of conveyor belts under the cages to remove manure, which provides better air control quality and eliminates fly breeding.

Original battery cages extended the technology used in battery brooders, which were cages with a wire mesh floor and integral heating elements for brooding chicks. The wire floor allowed the manure to pass through, removing it from the chicks' environment and reducing the risk of manure-borne diseases.

Early battery cages were often used for selecting hens based on performance, since it is easy to track how many eggs each hen is laying if only one hen is placed in a cage. Later, this was combined with artificial insemination, giving a technique where each egg's parentage is known. This method is still used today.

Early reports from Arndt about battery cages were enthusiastic. Arndt reported:

"This form of battery is coming into widespread use throughout the country and apparently is solving a number of the troubles encountered with laying hens in the regular laying house on the floor.

A chicken coop from the 1950s

In the first edition of this book I spoke of my experimental work with 220 pullets which were retained for one year in individual cages. At the end of this year it was found that the birds confined in the batteries outlaid considerably the same size flock in the regular houses. The birds consume less feed than those on the floor and this coupled with the increased production made them more profitable than the same number of pullets in the laying house.

A number of progressive poultrymen from all over the United States and some in foreign countries cooperated with me in carrying on experimental work with this type of battery and each and every one of them were very well satisfied with the results obtained. In fact, a number of them have since placed their entire laying flocks in individual hen batteries."

In 1967, Samuel Duff filed a patent for "battery cages" in patent US3465722.

The use of laying batteries increased gradually, becoming the dominant method somewhat before the integration of the egg industry in the 1960s. The practice of battery cages was criticized in Ruth Harrison's landmark book *Animal Machines*, published in 1964.

A simple battery cage system with no conveyors for feed or eggs

In 1990, North and Bell reported that 75% of all commercial layers in the world and 95% in the United States were kept in cages.

By all accounts, a caged layer facility is more expensive to build than high-density floor confinement, but can be cheaper to operate if designed to minimize labor.

North and Bell report the following economic advantages to laying cages:

1. It is easier to care for the pullets; no birds are underfoot. 2. Floor eggs are eliminated. 3. Eggs are cleaner. 4. Culling is expedited. 5. In most instances, less feed is required to produce a dozen eggs. 6. Broodiness is eliminated. 7. More pullets may be housed in a given house floor space. 8. Internal parasites are eliminated. 9. Labor requirements are generally much reduced

They also cite disadvantages to cages:

1. The handling of manure may be a problem. 2. Generally, flies become a greater nuisance. 3. The investment per pullet may be higher than in the case of floor operations. 4. There is a slightly higher percentage of blood spots in the eggs. 5. The bones are more fragile and processors often discount the fowl price.

Disadvantages 1 and 2 can be eliminated by manure conveyors, but some industrial systems do not feature manure conveyors.

## Legislation

### European Union

In 1999, the European Union Council Directive 1999/74/EC banned the conventional battery cage in the EU from 2012, after a 12-year phase-out. In their 1996 report, the European Commission's Scientific Veterinary Committee (SVC) condemned the battery cage, concluding:

"It is clear that because of its small size and its barrenness, the battery cage as used at present has inherent severe disadvantages for the welfare of hens".

The EU Directive allows enriched or "furnished" cages to be used. Under the directive, enriched cages must be at least 45 cm high and must provide each hen with at least 750 $cm^2$ of space; 600 $cm^2$ of this must be "usable area"– the other 150 $cm^2$ is for a nest-box. The cage must also contain litter, perches and "claw-shortening devices". Some animal welfare organisations, such as Compassion in World Farming, have criticised this move, calling for enriched cages to be prohibited as they believe they provide no significant or worthwhile welfare benefits compared with conventional battery cages.

Germany banned conventional battery cages from 2007, five years earlier than required by the EU Directive, and has prohibited enriched cages from 2012. Mahi Klosterhalfen of the Albert Schweitzer Foundation has been instrumental in a strategic campaign against battery cages in Germany.

## Switzerland

Switzerland banned battery cages from January 1, 1992; it was the first country to impose such a ban.

## United States

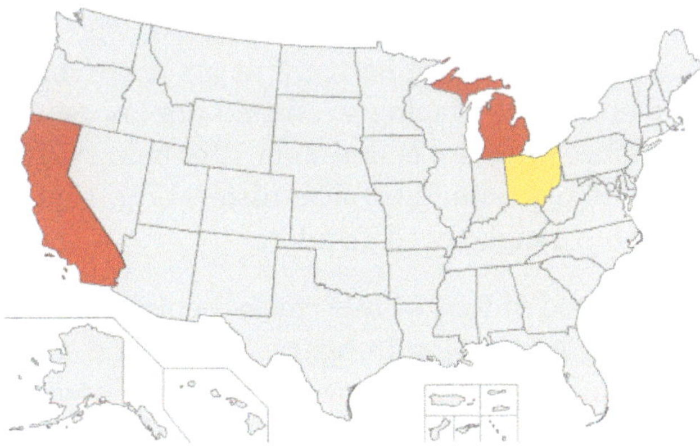

**US States with bans on battery cages for laying hens**

Laws prohibiting battery cages

Moratorium on permits for new battery cage construction

The passage of California Proposition 2 (2008) aimed, in part, to reduce or eliminate the problems associated with battery cages, by setting the standard for space relative to free movement and wingspan, rather than cage size.

Battery cages are also illegal in Michigan due to HB 5127, passed in 2009, which mandates that certain farm animals have enough room to stand up, lie down, turn around and extend their limbs, rather than being confined in tiny cages.

Finally, in Ohio, there is a moratorium on permits for the construction of new battery cages as of June 2010.

### Australia

The 2009 'Code of Practice' permits the use of battery cages. A written commitment by the Federal government to review the practice was scheduled in 2010; there was no further communication. During 2013 the state government of Tasmania was planning to phase out battery cages and budgeting for financial compensation for affected farmers but this was scrapped following the 2014 election. Attempts to change the law have been an object of contention; RSPCA Australia officially campaigns against cage eggs.

### Welfare Concerns

There are several welfare concerns regarding the battery cage system of housing and husbandry. These are presented below in the approximate chronological order they would influence the hens.

### Chick Culling

Due to modern selective breeding, laying hen strains are different from those of meat production strains. Male birds of the laying strains do not lay eggs and are unsuitable for meat production, therefore, they are culled soon after being sexed, often on the day of hatching. Methods of culling include cervical dislocation, asphyxiation by carbon dioxide and maceration using a high speed grinder.

Animal rights groups have used videos of live chicks being placed into macerators as evidence of cruelty in the egg production industry. Maceration, together with cervical dislocation and asphyxiation by carbon dioxide, are all considered acceptable methods of euthanasia by the American Veterinary Medical Association. Consumers may also be appalled simply by the death of animals that are not subsequently eaten.

### Beak-trimming

To reduce the harmful effects of feather pecking, cannibalism and vent pecking, most chicks eventually going into battery cages are beak-trimmed. This is often performed on the first day after hatching, simultaneously with sexing and receiving vaccinations. Beak-trimming is a procedure considered by many scientists to cause acute pain and distress with possible chronic pain; it is practised on chicks for all types of housing systems, not only battery cages.

## Cage Size

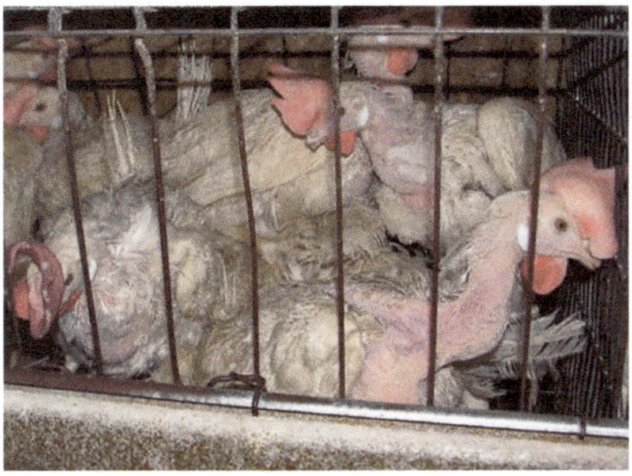

Battery cage

At approximately 16 weeks of age, pullets (hens which have not yet started to lay) are placed into cages. In countries with relevant legislation, floor space for battery cages ranges upwards from 300 cm² per bird. EU standards in 2003 called for at least 550 cm² per hen. In the US, the current recommendation by the United Egg Producers is 67 to 86 in² (430 to 560 cm²) per bird. The space available to each hen in a battery cage has often been described as less than the size of a sheet of A4 paper (624 cm²). Others have commented that a typical cage is about the size of a filing cabinet drawer and holds eight to 10 hens.

Behavioural studies showed that when turning, hens used 540 to 1006 cm², when stretching wings 653 to 1118 cm², when wing flapping 860 to 1980 cm², when feather ruffling 676 to 1604 cm², when preening 814 to 1240 cm², and when ground scratching 540 to 1005 cm². A space allowance of 550 cm² would prevent hens in battery cages from performing these behaviours without touching another hen. Animal welfare scientists have been critical of battery cages because of these space restrictions and it is widely considered that hens suffer boredom and frustration when unable to perform these behaviours. Spatial restriction can lead to a wide range of abnormal behaviours, some of which are injurious to the hens or their cagemates.

## Light Manipulation

To reduce the harmful effects of feather pecking, cannibalism and vent-pecking, hens in battery cages (and other housing systems) are often kept at low light intensities (e.g. less than 10 lux). Low light intensites may be associated with welfare costs to the hens as they prefer to eat in brightly lit environments and prefer brightly lit areas for active behaviour but dim (less than 10 lux) for inactive behaviour. Dimming the lights can also cause problems when the intensity is then abruptly increased temporarily to inspect the hens; this has been associated as a risk factor of increased feather pecking and the birds can become frightened resulting in panic-type ("hysteria") reactions which can increase the risk of injury.

Battery cages - note the low light intensity beyond range of the camera flashgun

Being indoors, hens in battery cages do not see sunlight. Whilst there is no scientific evidence for this being a welfare problem, some animal advocates indicate it is a concern. Furnished cages and some other non-cage indoor systems would also prevent hens seeing natural light throughout their lives.

## Osteoporosis

Several studies have indicated that toward the end of the laying phase (approximately 72 weeks of age), a combination of high calcium demand for egg production and a lack of exercise can lead to osteoporosis. This can occur in all housing systems for egg laying hens, but is particularly prevalent in battery cage systems where it has sometimes been called 'cage layer osteoporosis'. Osteoporosis leads to the skeleton becoming fragile and an increased risk of bone breakage, particularly in the legs and keel bone. Fractures may occur whilst the hens are in the cage and these are usually discovered at depopulation as old, healed breaks, or they might be fresh breaks which occurred during the process of depopulation. One study showed that 24.6% of hens from battery cages had recent keel fractures whereas hens in furnished cages, barn and free-range had 3.6%, 1.2% and 1.3% respectively. However, hens from battery cages experienced fewer old breaks (17.7%) compared to hens in barn (69.1%), free-range (59.8%) and furnished cages (31.7%).

## Forced Moulting

Flocks are sometimes force moulted, rather than being slaughtered, to reinvigorate egg-laying. This involves complete withdrawal of food (and sometimes water) for 7 to 14 days or sufficiently long to cause a body weight loss of 25 to 35%. This stimulates the hen to lose her feathers, but also reinvigorates egg-production. Some flocks may be force moulted several times. In 2003, more than 75% of all flocks were moulted in the US. This temporary starving of the hens is seen as inhumane and is the main point of objection by critics and opponents of the practice. The alternative most often employed is to slaughter the hens instead of moulting them.

### Improving Welfare for Egg Producing Hens

The Scientific Veterinary Committee of the European Commission stated that "enriched cages and well designed non-cage systems have already been shown to have a number of welfare advantages over battery systems in their present form". Supporters of battery husbandry contend that alternative systems such as free range also have welfare problems, such as increases in cannibalism, feather pecking and vent pecking. A recent review of welfare in battery cages made the point that such welfare issues are problems of management, unlike the issues of behavioural deprivation, which are inherent in a system that keeps hens in such cramped and barren conditions. Free range egg producers can limit or eliminate injurious pecking, particularly feather pecking, through such strategies as providing environmental enrichment, feeding mash instead of pellets, keeping roosters in with the hens, and arranging nest boxes so hens are not exposed to each other's vents; similar strategies are more restricted or impossible in battery cages.

## Furnished Cages

Furnished cages, sometimes called "enriched" or "modified" cages, are cages for egg laying hens which have been designed to overcome some of the welfare concerns of battery cages whilst retaining their economic and husbandry advantages, and also provide some of the welfare advantages of non-cage systems. Many design features of furnished cages have been incorporated because research in animal welfare science has shown them to be of benefit to the hens.

### History and Legislation

Battery cages are already banned in several countries including Switzerland, Belgium, Austria, Sweden and the Netherlands and prototype commercial furnished cage systems were being developed in the 1980s. In 1999, the European Union Council Directive 1999/74/EC banned the conventional battery cage in the EU from 2012, after a 10-year phase-out. As alternatives to battery cages, the EU Council Directive allowed non-cage systems and furnished cages. Furnished cages therefore represent a feasible alternative to battery cages in the EU after 2012.

Under Directive 1999/74/EC, furnished cages must provide at least the following: 750 $cm^2$ per hen, of which 600 $cm^2$ is 45 cm high, a nest, a littered area for scratching and pecking, 15 cm of perch and 12 cm of food trough per hen and a claw shortening device.

Austria banned battery cages in 2009 and is set to ban furnished cages by 2020. Belgium has also banned the battery cage – and proposes to ban furnished cages by 2024. Germany has introduced a 'family cage', which has more space than the furnished cages used in other countries, however, consumers in Germany have been rejecting these eggs. Outside the EU, Switzerland has already banned both the battery and furnished cage systems.

All major UK supermarkets have promised to stop selling eggs from furnished cages by 2025.

## Furnished Cages and Battery Cages

Furnished cages retain several advantages of battery cages in that they-

- Separate the eggs from the hens' feces thereby keeping the eggs clean
- Protect the hens from predation
- Automatically collect the eggs thereby preventing egg-eating and floor-laying which both incur additional cost
- Retain a small group size which reduces injurious pecking behaviour

Furnished cages have welfare benefits additional to battery cages by providing -

- Additional space
- A nest
- A claw shortening device
- A dust bath/litter substrate
- A perch
- Easier access for depopulation

## Current Designs

There is no clear limit to the size of the furnished cages. Although initial models were not much larger than conventional battery cages, most current designs house 40 to 80 hens although one system houses 115 hens. The depth of furnished cages is often more than the depth of battery cages and as a result, they are often arranged with only one cage row per level, i.e. not connected back-to-back. The more shallow cages can be connected back-to-back. To create space for large groups of hens, some designs of furnished cages are very long. Cage bottoms are made of wire mesh or plastic slats and are sloped so that eggs not laid in the nest box roll onto an egg belt. Feed is provided in feeders outside the cage, although in some designs there may be internal feeders or a combination of the two. Perches in some designs are raised and in others are at floor level.

## Welfare Benefits

In a study which compared the welfare benefits of hens in furnished cages, battery cages, free range and barn systems, hens in furnished cages had the lowest faecal corticosterone (a hormone that indicates stress levels), the lowest number of hens that were vent pecked, lowest number of egg shells with calcium spots (an indicator of stress when the egg is temporarily retained by the hen), lowest number of egg shells with blood spots on (usually caused by prolapse), lowest score of skin damage, lowest severity of vent damage caused by vent pecking and lowest plumage soiling. Hens in furnished cages had a similar percentage of hens with recent keel fractures which are usually caused during depopulation (3.6%) compared to hens in barn (1.2%) and free-range systems (1.3%), all of which were considerably lower than in hens from battery cages (24.6%). Furthermore, hens in furnished cages had a smaller percentage of old keel fractures (31.7%) compared to hens in barn (69.1%) and free-range (59.8%) systems but more than hens in battery cages (17.7%). This indicates that furnished cages protect against the keel breaks that are common amongst non-caged hens and also protects against the effects of osteoporosis prevalent in battery cages causing bones to be weak and easily broken during depopulation. In this study, mortality rates were above the breed standards in all systems except the furnished cages.

### Welfare Disadvantages

Furnished cages provide more space than battery cages but still prevent some behaviours such as vigorous wing-flapping, flying, nest-building (no materials are provided) and inhibit others (comfort or grooming behaviours) determined partly by the numbers of hens in the cage. The hens are not separated from their feces as completely as hens in battery cages and therefore are at a greater risk of disease, although not as great as the risk to hens in non-cage systems. The small amount of litter that is provided in furnished cages is often distributed quickly or flicked out the cage, possibly resulting in frustration for hens wishing to dustbath and resulting in sham dustbathing. The nest boxes are often occupied by hens using the box for behaviours other than egg-laying (e.g. for sleeping or sham dustbathing) which could lead to frustration in hens wishing to lay an egg.

### Production in Furnished Cages

Some studies indicate that production in furnished cages is comparable to that in battery cages. Other studies indicate hens housed in furnished cages have better bodyweights and egg production compared to hens in battery cages.

## Debeaking

Debeaking, also called beak trimming, is the partial removal of the beak of poultry, especially layer hens and turkeys although it may also be performed on quail and ducks. Most commonly, the beak is shortened permanently, although regrowth can occur. The trimmed lower beak is somewhat longer than the upper beak. USA's UEP guidelines suggest that in egg laying strains of chickens, the length of the upper beak distal from the nostrils that remains following trimming, should be 2 to 3 mm. In the UK, the Farm Animal Welfare Council stated: "The accepted procedure is to remove not more than one third of the upper and lower beaks or not more than one third of the upper beak only" but went on to recommend: "Where beak trimming is carried out, it should, wherever possible, be restricted to beak tipping; that is the blunting of the beak to remove the sharp point which can be the cause of the most severe damage to other birds." The term "debeaking" implies that the entire beak is removed during the trimming process, though in reality only half or less of the beak is removed.

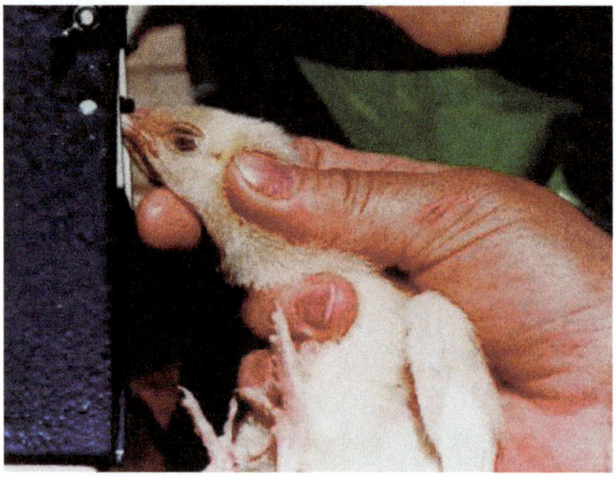

A chick being beak trimmed with a heated-blade device.

Beak trimming is most common in egg-laying strains of chickens. In some countries such as the United States, turkeys are routinely beak trimmed, however, in the UK, only 10% of turkeys are beak trimmed. Beak trimming is a preventive measure to reduce damage caused by injurious pecking such as cannibalism, feather pecking and vent pecking, and thereby improve livability. Commercial broiler chickens are not routinely beak trimmed as they reach slaughter weight at approximately 6 weeks of age, i.e. before injurious pecking usually begins. However, broiler breeding stock may be trimmed to prevent damage during mating. In some countries, beak trimming is done as a last resort where alternatives are considered not to be possible or appropriate.

Beak trimming has been banned in Switzerland since 1992 and will be phased out in Germany by 2017.

In close confinement, cannibalism, feather pecking and aggression are common among turkeys, ducks, pheasants, quail, and egg laying strains of chickens of many breeds (including both heritage breeds and modern hybrids). The tendency to cannibalism and feather pecking varies among different strains of chickens, but does not manifest itself consistently. Some flocks of the same breed may be entirely free from cannibalism, while others, under the same management, may have a serious outbreak. Mortalities mainly due to cannibalism can be up to 15% in egg laying flocks housed in aviaries, straw yards, and free-range systems.

An adult bird which has been beak-trimmed as a chick

Because egg laying strains of chickens can be kept in smaller group sizes in caged systems, cannibalism is reduced leading to a lowered trend in mortality as compared to non-cage systems. Cannibalism among flocks is highly variable and when it is not problematic, then mortalities among production systems are similar.

## History

Beak trimming was developed at the Ohio Experiment Station in the 1930s. The original technique was temporary, cutting approximately 6 mm (1/4 inch) off the beak. It was thought that the tip of the beak had no blood supply and presumably no sensation. The procedure was performed by hand with a sharp knife, either when deaths due to cannibalism became excessive, or when the problem was anticipated because of a history of cannibalism in the particular strain of chicken.

Cannibalism is a serious management problem dating back to the periods before intensive housing of poultry became popular. Poultry books written before vertical integration of the poultry industry describe the abnormal pecking of poultry:

Chicks and adult birds' picking at each other until blood shows and then destroying one another by further picking is a source of great loss in many flocks, especially when kept in confinement .... The recommendation of the Ohio Experiment Station of cutting back the tip of the upper beak has been found to be effective until the beak grows out again.

Cannibalism has two peaks in the life of a chicken; during the brooding period and at the onset of egg laying. The point-of-lay cannibalism is generally the most damaging and gets most of the attention. The temporary beak trimming developed at the Ohio Experiment Station assumed that cannibalism was a phase, and that blunting the beak temporarily would be adequate.

## Current Methods

In recent years, the aim has been to develop more permanent beak trimming (although repeat trimming may be required), using electrically heated blades in a beak trimming machine, to provide a self-cauterizing cut. There are currently (2012) four widely used methods of beak trimming: hot blade, cold blade (including scissors or secateurs), electrical (the Bio-beaker) and infrared. The latter two methods usually remove only the tip of the beak and do not leave an open wound, therefore they may offer improvements in welfare. Other approaches such as the use of lasers, freeze drying and chemical retardation have been investigated but are not in widespread use. The infrared method directs a strong source of heat into the inner tissue of the beak and after a few weeks, the tip of the upper and lower beak dies and drops off making the beak shorter with blunt tips. The Bio-beaker, which uses an electric current to burn a small hole in the upper beak, is the preferred method for trimming the beaks of turkeys. The Farm Animal Welfare Council (FAWC) wrote regarding beak trimming of turkeys that cold cutting was the most accurate method, but that substantial re-growth of the beak occurred; although the Bio-beaker limited beak re-growth, it was less accurate. It was considered that the hot cut was the most distressing procedure for turkeys.

In the UK, beak trimming of layer hens normally occurs at 1-day of age at the same time as the chick is being sexed and vaccinated.

## Legislation

### Australia

Debeaking is banned in the Australian Capital Territory. It is currently debated in parliament whether debeaking should be banned in Victoria and New South Wales.

### Denmark

Debeaking was used in Denmark until 2014 when it was phased out and made illegal.

### Sweden, Finland and Norway

Debeaking is illegal in these countries.

## Costs and Benefits

### Costs

The costs of beak trimming relate primarily to welfare concerns. These include acute stress, and acute, possibly chronic, pain following trimming. A bird's ability to consume feed is impaired following beak trimming because of the new beak shape and pain. Most studies report reduced body weights and feed intake following beak trimming, however, by sexual maturity or peak egg production, growth rates are usually normal. Weight losses were reduced in chicks that were beak trimmed by infrared compared with chicks trimmed by a hot-blade.

## The Pain of Beak Trimming

## White Leghorn Pullets Showing the Results of Beak Trimming

Non-beak trimmed

Whether beak trimming causes pain is a hotly debated concern. It is a complex issue as it may involve acute and/or chronic pain, and depends on the age it is performed, the method of trimming and the length of beak that is removed. Beak trimming in the poultry industry usually occurs without anaesthetic at 1-day of age or when the chicks are very young, but can occur at a later age if an outbreak of feather pecking occurs, and in some cases, birds may be beak trimmed on repeated occasions. Beak trimming is not permitted in the UK on meat chickens that are aged over 10 days.

Beak trimmed

## Acute Pain

The beak is a complex, functional organ with an extensive nervous supply including nociceptors that sense pain and noxious stimuli. These would almost certainly be stimulated during beak trimming, indicating strongly that acute pain would be experienced. Behavioural evidence of pain after beak trimming in layer hen chicks has been based on the observed reduction in pecking behavior, reduced activity and social behavior, and increased sleep duration. In Japanese quail, beak-trimming by cauterization caused lower body weights and feed intake in the period just after beak trimming. Beak trimmed Muscovy ducks spent less time engaging in beak-related behaviours (preening, feeding, drinking, exploratory pecking) and more time resting than non-trimmed ducks in the days immediately post-trim. These differences disappeared by 1 week post-trim. At 1 week post-trim the trimmed ducks weighed less than non-trimmed ducks, but this difference disappeared by 2 weeks post-trim. It is, however, unclear if the above changes in behaviour arise from pain or from a loss of sensitivity in the beak. Pecking force has been found to decrease after beak trimming in adult hens possibly indicating that hens are protecting a painful area from further stimulation. However, pecking force did not differ between chicks with or without minor beak-trims at 2 to 9 days of age, suggesting that chicks with minor beak-trims do not experience pain from the beak.

## Chronic Pain

Severe beak trimming, or beak trimming birds at an older age is thought to cause chronic pain. Following beak trimming of older or adult hens, the nociceptors in the beak stump show abnormal patterns of neural discharge, which indicate acute pain. Neuromas, tangled masses of swollen regenerating axon sprouts, are found in the healed stumps of birds beak trimmed at 5 weeks of age or older and in severely beak trimmed birds. Neuromas have been associated with phantom pain in human amputees and have therefore been linked to chronic pain in beak trimmed birds. If beak trimming is severe because of improper procedure or done in older birds, the neuromas will persist which suggests that beak trimmed older birds experience chronic pain, although this has been debated.

## Benefits

The benefits of beak trimming are mainly welfare advantages for birds kept in close confinement, some of which directly relate to increases (or reduced decreases) in production. These include reduced feather pecking and cannibalism, better feather condition, less fearfulness and nervousness, less chronic stress, and decreased mortality.

## Alternatives

A range of options have been proposed as possible alternatives to beak trimming including modifying the genetics of domesticated poultry to reduce cannibalistic tendencies. For confined housing where light control is possible, lowering light intensity so that birds cannot see each other as easily reduces antagonistic encounters and aggressive behaviour. Enrichment devices, introduced at an early age, as simple objects hung in a habitat can reduce aggressive behavior. Dividing the population into smaller group sizes reduces cannibalism. Proper body weight management that avoids underweight pullets reduces the probability of underweight pullets with uterine prolapse that leads to cloacal cannibalism.

## Chick Culling

Chick culling is the process of killing newly hatched poultry for which the industry has no use. Due to modern selective breeding, laying hen strains differ from meat production strains. As male birds of the laying strain do not lay eggs, they are generally killed soon after they hatch and shortly after being sexed. Methods of culling include cervical dislocation, asphyxiation by carbon dioxide and maceration using a high speed grinder.

### History

Prior to the development of modern broiler meat breeds, most male chickens (cockerels) were slaughtered for meat, whereas females (pullets) would be kept for egg production. However, once the industry bred separate meat and egg-producing hybrids, there was no reason to keep males of the egg-producing hybrid. As a consequence, the males of egg-laying chickens are killed as soon as possible after hatching and sexing to reduce losses incurred by the breeder. Special techniques have been developed to accurately determine the sex of chicks at as young an age as possible.

It has been reported that in India for example, more than 180 million male chicks per year are culled. The egg industry in India is growing at the rate of 8-12% yearly, and is the third largest egg producer.

Chicks are also culled in the production of foie gras. After hatching, the ducklings are sexed. Males put on more weight than females, so the females are killed, sometimes in an industrial macerator. Up to 40 million female ducks per year may be killed in this way. The remains of female ducklings are later used in cat food, fertilisers and in the pharmaceutical industry.

### Methods

Several methods can be used to kill chicks:

- Maceration, using a large high-speed grinder into which the live chicks are fed.
- Gases or gas mixtures, often carbon dioxide is used to induce unconsciousness

and then death.

- Cervical dislocation (breaking the neck)
- Electrocution, a new method that has been touted as being cheap, reliable, and humane by its developers
- Suffocation in plastic bags

Chick grinding machine

## US Recommended Methods

The American Veterinary Medical Association recommends cervical dislocation, maceration, and asphyxiation by carbon dioxide as the better options.

The 2005-2006 American Veterinary Medical Association Executive Board proposed a policy change, which was recommended by the Animal Welfare Committee on disposal of unwanted chicks, poults, and pipped eggs. The policy states "Unwanted chicks, poults, and pipped eggs should be killed by an acceptable humane method, such as use of a commercially designed macerator that results in instantaneous death. Smothering unwanted chicks or poults in bags or containers is not acceptable. Pips, unwanted chicks, or poults should be killed prior to disposal. A pipped egg, or pip, is one where the chick or poult has not been successful in escaping the egg shell during the hatching process."

## Controversy

Animal rights and animal welfare advocates maintain that many of the current practices surrounding chicken slaughtering are unethical.

## Alternatives

A Unilever spokesperson has been quoted as saying "We have also committed to providing funding and expertise for research and introduction of alternative methods such as in-ovo gender identification (sexing) of eggs. This new technology offers the potential to eliminate the hatching and culling of male chicks."

## Forced Molting

Forced molting typically involves the removal of food and/or water from poultry for an extended period of time to reinvigorate egg-laying.

Forced molting, sometimes known as induced molting, is the practice by some poultry industries of artificially provoking a flock to molt simultaneously, typically by withdrawing food for 7–14 days and sometimes also withdrawing water for an extended period. Forced molting is usually implemented when egg-production is naturally decreasing toward the end of the first egg-laying phase. During the forced molt, the birds cease producing eggs for at least two weeks, which allows the bird's reproductive tracts to regress and rejuvenate. After the molt, the hen's egg production rate usually peaks slightly lower than the previous peak, but egg quality is improved. The purpose of forced molting is therefore to increase egg production, egg quality, and profitability of flocks in their second or subsequent laying phases.

The practice is controversial. While it is widespread in the US, it is prohibited in the EU.

### Background

Commercial hens usually begin laying eggs at 16–20 weeks of age, although production gradually declines soon after from approximately 25 weeks of age. This means that in many countries, by approximately 72 weeks of age, flocks are considered economically unviable and are slaughtered after approximately 12 months of egg production, although chickens will naturally live for 6 or more years. However, in some countries, rather than being slaughtered, the hens are force molted to re-invigorate egg-laying for a second, and sometimes subsequent, laying phase.

Forced molting simulates the natural process where chickens grow a new set of feathers in the Autumn, a process generally accompanied by a sharp reduction or cessation of egg production. Natural molting is stimulated by shortening day lengths combined with stress (of any kind). Before confinement housing with artificial lights was the norm, the Autumn molt caused a seasonal scarcity of eggs and high market prices. Farmers attempted to pamper their flocks to prevent the molt as long as possible, to take advantage of the high prices. Modern controlled-environment confinement housing has the opposite problem; the hens are not normally presented with sufficient stress or cues to go into molt naturally. However, after laying almost daily for nearly a year, their rate of egg production declines, as does the quality of the eggshell and the egg contents. In addition, the hens are overweight.

It is sometimes claimed that forced molting is an artifact of modern intensive farming, but the practice predates the vertical integration of the poultry industry by decades; former Head of the Poultry Science Department at the University of Maryland, Morley A. Jull prescribes a precise molting program in his 1938 book, *Poultry Husbandry*.

## Methods

For a complete recovery of the reproductive tract, the hen's body weight must drop by 30 to 35 percent during the forced molt. This is typically achieved by withdrawing the hen's feed for 7–14 days, sometimes up to 28 days. This induces the birds to lose their feathers, cease to lay eggs and lose body-weight. Some programs combine feed withdrawal with a period of water withdrawal. Most programs also restrict the amount of lighting to provide a daylight period that is too short to stimulate egg production, providing a simulated autumn, the natural time of molt and minimum egg production.

Forced molting programs sometimes follow other variations. Some do not eliminate feed altogether, but may induce a molt by providing a low-density diet (e.g. grape pomace, cotton seed meal, alfalfa meal) or dietary manipulation to create an imbalance of a particular nutrient(s). The most important among these include manipulation of minerals including sodium, calcium, iodine and zinc, with full or partially reduced dietary intakes. These alternative methods of forced molting have not been widely used by the egg industries.

In 2003, more than 75% of all flocks in the US were molted.

## Mortality

Some birds die during forced molting and it has been recommended that the flock must be managed so that mortality does not exceed 1.25% over the 1–2 weeks of (nearly complete) feed withdrawal, compared to a 0.5% to 1.0% monthly mortality in a well-managed flock under low-stress conditions. Alternative methods of forced molting which do not use total food withdrawal, e.g. creating a dietary mineral imbalance, generally result in lower mortality rates.

### Controversy

Temporary starvation of the hens is considered by some to be inhumane and is the main objection of critics and opponents of the practice.

While forced molting is widespread in the US, it is prohibited in the EU. In the UK, the Department for Environment, Food and Rural Affairs (Defra) states *In no circumstances may birds be induced to moult by withholding feed and water*. Forced molting is not a common practice in Canada, where the animal welfare issues associated with it have rendered it basically obsolete.

Forced molting increases plasma corticosterone which, along with related hormones, decreases the levels of circulating lymphocytes and other leukocytes, thereby reducing the effectiveness of the bird's immune system. This means that birds being force molted become susceptible to disease, particularly *Salmonella* infections, and may produce contaminated eggs thereby compromising public health.

## Pastured Poultry

Pastured poultry is a sustainable agriculture technique that calls for the raising of laying chickens, meat chickens (broilers), and/or turkeys on pasture, as opposed to indoor confinement. Humane treatment, the perceived health benefits of pastured poultry, and superior texture and flavor, are causing an increase in demand for such products.

Joel Salatin of Swoope, Virginia, helped to reintroduce the technique at Polyface Farm, and wrote his book *Pastured Poultry Profits* to spread the idea to other farmers. Andy Lee and Herman Beck-Chenoweth expanded on Salatin's techniques, and created some of their own.

The American Pastured Poultry Producers' Association (APPPA) was formed to promote pastured poultry. Its membership consists largely of pastured poultry farmers.

Though pasture feeding improves the nutritive quality of ruminant meats, the effect of pasture feeding on poultry meat composition is not well established. One trial showed low impact of pasture feeding on vitamin E and fatty acid composition.

The pens that house the fowl can be made from wood and scrap metal or out of PVC pipe and white tarps.

Pastured poultry is also gaining popularity because it helps the farmer, through reducing capital costs, and increasing pasture fertility. It is very well suited for incorporation within a system of managed intensive grazing.

Pastured poultry is not limited to chickens and turkeys. It includes a variety of other birds, including ducks, geese, and exotics in the poultry family.

## Free-range Poultry

A chicken tractor in use as part of a pastured poultry system

Herman Beck-Chenoweth reintroduced the free-range system that was the most popular way to raise poultry in the U.S. from the 1930s through the 1960s. The system allows birds to range freely during the day and be safely sequestered on secure skid houses over night. The addition of a guard animal, such as an Komondor or Anatolian Shepherd dog, controls predators. In the modern American free-range poultry production system, birds are much less crowded and freer to practice normal bird behaviour than in any other pasture-based system.

Although frequently listed as a "pasture" method, free-"range" refers to the length of the forage. Cattle graze "pasture" which is forage over six inches long. "Range" refers to short forage of 2-4 inches. Free-range is a very sustainable production system that improves the farmer's soil and produces poultry with strong bones and meat with good "mouthfeel". Combined with proper aging after slaughter, the meat is tender and flavorful.

## References


- North, Mack O.; Donald E. Bell (1990). Commercial Chicken Production Manual (4th ed.). Van Nostrand Reinhold. pp. 297, 315. ISBN 0-87055-446-8.
- Appleby, M.C.; Mench, J.A.; Hughes, B.O. (2004). Poultry Behaviour and Welfare. Wallingford and Cambridge MA: CABI Publishing. ISBN 0-85199-667-1.
- "Landmark Ohio Animal Welfare Agreement Reached Among HSUS, Ohioans for Humane Farms, Gov. Strickland, and Leading Livestock Organizations". Humane Society of the United States. 30 June 2010. Retrieved 2 March 2015.
- AVMA (2010). "AVMA Literature review on the welfare implications of beaktrimming" (PDF). Retrieved November 23, 2015.

- Saul, H. (March 5, 2015). “Hatched, discarded, gassed: What happens to male chicks in the UK”. Retrieved July 1, 2015.
- Saraswathy, M. “Unilever working to end the culling of male chicks”. Business Standard. Retrieved February 5, 2015.
- Hughes, I. (2014). “Shocking video shows hundreds of live ducklings ‘thrown into mincer’ on cruel ‘foie gras farm’”. The Mirror. Retrieved March 14, 2015.
- Bruce Friedrich (14 January 2013). “The Cruelest of All Factory Farm Products: Eggs From Caged Hens”. Retrieved 2 March 2015.
- “Innovax-Production in Furnished Cages ii”. Archived from the original on April 25, 2012. Retrieved 15 November 2011.

5

# Various Livestock Breeds in Poultry

Various breeds of livestock are used in poultry. Some of these birds are Andalusian chicken, Bresse Gauloise, Delaware chicken, Asil chicken, broiler and Indian Runner ducks. This section has been carefully written to provide an easy understanding of the numerous kinds of livestock in poultry.

## Chicken

The chicken (*Gallus gallus domesticus*) is a type of domesticated fowl, a subspecies of the red junglefowl. It is one of the most common and widespread domestic animals, with a population of more than 19 billion as of 2011. Humans keep chickens primarily as a source of food, consuming both their meat and their eggs.

Genetic studies have pointed to multiple maternal origins in Southeast-, East-, and South Asia, but with the clade found in the Americas, Europe, the Middle East and Africa originating in the Indian subcontinent. From India, the domesticated chicken was imported to Lydia in western Asia Minor, and to Greece by the fifth century BC. Fowl had been known in Egypt since the mid-15th century BC, with the "bird that gives birth every day" having come to Egypt from the land between Syria and Shinar, Babylonia, according to the annals of Thutmose III. The chicken genome has changed less from feathered ancestors eradicated by the Cretaceous–Paleogene extinction event than those of other sequenced avian dinosaurs.

### Terminology

In the UK and Ireland adult male chickens over the age of one year are primarily known as cocks, whereas in America, Australia and Canada they are more commonly called roosters. Males less than a year old are *cockerels*. Castrated roosters are called *capons* (surgical and chemical castration are now illegal in some parts of the world). Females over a year old are known as *hens* and younger females as *pullets* although in the egg-laying industry, a pullet becomes a hen when she begins to lay eggs at 16 to 20 weeks of age. In Australia and New Zealand (also sometimes in Britain), there is a generic term *chook* /□t□□k/ to describe all ages and both sexes. The young are called *chicks* and the meat is called *chicken*.

"Chicken" originally referred to young domestic fowl. The species as a whole was then called *domestic fowl*, or just *fowl*. This use of "chicken" survives in the phrase "Hen and Chickens", sometimes used as a British public house or theatre name, and to name groups of one large and many small rocks or islands in the sea (for example Hen and Chicken Islands). The word "chicken" is sometimes erroneously construed to mean females exclusively, despite the term "hen" for females being in wide circulation.

In the Deep South of the United States chickens are also referred to by the slang term *yardbird*.

## General Biology and Habitat

Chickens are omnivores. In the wild, they often scratch at the soil to search for seeds, insects and even animals as large as lizards, small snakes or young mice.

Chickens may live for five to ten years, depending on the breed. The world's oldest chicken, a hen, died of heart failure at the age of 16 according to Guinness World Records.

Roosters can usually be differentiated from hens by their striking plumage of long flowing tails and shiny, pointed feathers on their necks (*hackles*) and backs (*saddle*), which are typically of brighter, bolder colours than those of females of the same breed. However, in some breeds, such as the Sebright chicken, the rooster has only slightly pointed neck feathers, the same colour as the hen's. The identification can be made by looking at the comb, or eventually from the development of spurs on the male's legs (in a few breeds and in certain hybrids, the male and female chicks may be differentiated by colour). Adult chickens have a fleshy crest on their heads called a comb, or cockscomb, and hanging flaps of skin either side under their beaks called wattles. Collectively, these and other fleshy protuberances on the head and throat are called caruncles. Both the adult male and female have wattles and combs, but in most breeds these are more prominent in males. A *muff* or *beard* is a mutation found in several chicken breeds which causes extra feathering under the chicken's face, giving the appearance of a beard. Domestic chickens are not capable of long distance flight, although lighter birds are generally capable of flying for short distances, such as over fences or into trees (where they would naturally roost). Chickens may occasionally fly briefly to explore their surroundings, but generally do so only to flee perceived danger.

## Behaviour

Hen with chicks, India

## Social Behaviour

Chickens are gregarious birds and live together in flocks. They have a communal approach to the incubation of eggs and raising of young. Individual chickens in a flock will dominate others, establishing a "pecking order", with dominant individuals having priority for food access and nesting locations. Removing hens or roosters from a flock causes a temporary disruption to this social order until a new pecking order is established. Adding hens, especially younger birds, to an existing flock can lead to fighting and injury. When a rooster finds food, he may call other chickens to eat first. He does this by clucking in a high pitch as well as picking up and dropping the food. This behaviour may also be observed in mother hens to call their chicks and encourage them to eat.

A rooster's crowing is a loud and sometimes shrill call and sends a territorial signal to other roosters. However, roosters may also crow in response to sudden disturbances within their surroundings. Hens cluck loudly after laying an egg, and also to call their chicks. Chickens also give a low "warning call" when they think they see a predator approaching.

## Courtship

To initiate courting, some roosters may dance in a circle around or near a hen ("a circle dance"), often lowering his wing which is closest to the hen. The dance triggers a response in the hen and when she responds to his "call", the rooster may mount the hen and proceed with the mating.

## Nesting and Laying Behaviour

Chicken eggs vary in colour depending on the hen, typically ranging from bright white to shades of brown and even blue, green, and recently reported purple (found in South Asia) (Araucana varieties).

Hens will often try to lay in nests that already contain eggs and have been known to move eggs from neighbouring nests into their own. The result of this behaviour is that a flock will use only a few preferred locations, rather than having a different nest for every bird. Hens will often express a preference to lay in the same location. It is not unknown for two (or more) hens to try to share the same nest at the same time. If the nest is small, or one of the hens is particularly determined, this may result in chickens trying to lay on top of each other. There is evidence that individual hens prefer to be either solitary or gregarious nesters. Some farmers use fake eggs made from plastic or stone (or golf balls) to encourage hens to lay in a particular location.

Chicks before their first outing

## Broodiness

Under natural conditions, most birds lay only until a clutch is complete, and they will then incubate all the eggs. Many domestic hens will also do this–and are then said to "go broody". The broody hen will stop laying and instead will focus on the incubation of the eggs (a full clutch is usually about 12 eggs). She will "sit" or "set" on the nest, protesting or pecking in defense if disturbed or removed, and she will rarely leave the nest to eat, drink, or dust-bathe. While brooding, the hen maintains the nest at a constant temperature and humidity, as well as turning the eggs regularly during the first part of the incubation. To stimulate broodiness, an owner may place many artificial eggs in the nest, or to stop it they may place the hen in an elevated cage with an open wire floor.

Modern egg-laying breeds rarely go broody, and those that do often stop part-way through the incubation. However, some "utility" (general purpose) breeds, such as the Cochin, Cornish and Silkie, do regularly go broody, and they make excellent mothers, not only for chicken eggs but also for those of other species—even those with much smaller or larger eggs and different incubation periods, such as quail, pheasants, turkeys or geese. Chicken eggs can also be hatched under a broody duck, with varied success.

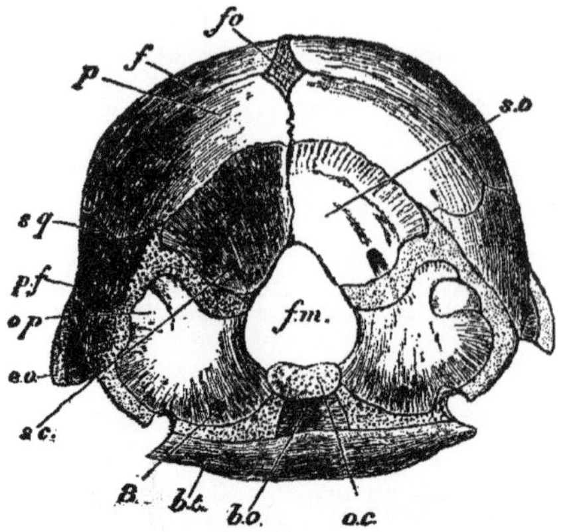


Skull of a three-week-old chicken. Here the opisthotic bone appears in the occipital region, as in the adult Chelonian. bo = Basi-occipital, bt = Basi-temporal, eo = Opisthotic, f = Frontal, fm = Foramen magnum, fo = Fontanella, oc = Occipital condyle, op = Opisthotic, p = Parietal, pf = Post-frontal, sc = Sinus canal in supra-occipital, so = Supra-occpital, sq = Squamosal, 8 = Exit of vagus nerve.

## Hatching and Early Life

At the end of the incubation period (about 21 days), the eggs, if fertile, will hatch. Development of the egg starts only when incubation begins, so they all hatch within a day or two of each other, despite perhaps being laid over a period of two weeks or so. Before hatching, the hen can hear the chicks peeping inside the eggs, and will gently cluck to stimulate them to break out of their shells. The chick begins by "pipping"; pecking a breathing hole with its egg tooth towards the blunt end of the egg, usually on the upper side. The chick will then rest for some hours, absorbing the remaining egg yolk and withdrawing the blood supply from the membrane beneath the shell (used earlier for breathing through the shell). It then enlarges the hole, gradually turning round as it goes, and eventually severing the blunt end of the shell completely to make a lid. It crawls out of the remaining shell, and its wet down dries out in the warmth of the nest.

The hen will usually stay on the nest for about two days after the first egg hatches, and during this time the newly hatched chicks live off the egg yolk they absorb just before hatching. Any eggs not fertilized by a rooster will not hatch, and the hen eventually loses interest in these and leaves the nest. Sometimes the hen will eat the eggs, as a matter of habit. After hatching, the hen fiercely guards the chicks, and will brood them when necessary to keep them warm, at first often returning to the nest at night. She leads them to food and water; she will call them to edible items, but seldom feeds them directly. She continues to care for them until they are several weeks old, when she will gradually lose interest and eventually start to lay again.

## Embryology

In 2006, scientists researching the ancestry of birds "turned on" a chicken recessive gene, *talpid2*, and found that the embryo jaws initiated formation of teeth, like those found in ancient bird fossils. John Fallon, the overseer of the project, stated that chickens have "...retained the ability to make teeth, under certain conditions... ."

## Breeding

### Origins

Male red junglefowl.

The domestic chicken is descended primarily from the red junglefowl (*Gallus gallus*) and is scientifically classified as the same species. As such it can and does freely interbreed with populations of red jungle fowl. Recent genetic analysis has revealed that at least the gene for yellow skin was incorporated into domestic birds through hybridization with the grey junglefowl (*G. sonneratii*). The traditional poultry farming view is that chickens were first domesticated for cockfighting in Asia, Africa, and Europe, rather than for egg or meat production. In the last decade there have been a number of genetic studies to clarify the origins. According to one study, a single domestication event occurring in the region of modern Thailand created the modern chicken with minor transitions separating the modern breeds. However, that study was later found to be based on incomplete data, and recent studies point to multiple maternal origins, with the clade found in the Americas, Europe, Middle East, and Africa, originating from the Indian subcontinent, where a large number of unique haplotypes occur. It is postulated that the jungle fowl, known as the bamboo fowl in many Southeast Asian languages, is a special pheasant well adapted to take advantage of the large amounts of fruits that are produced during the end of the 50-year bamboo seeding cycle to boost its own reproduction. In domesticating the chicken, humans took advantage of this prolific reproduction of the jungle fowl when exposed to large amounts of food.

It has been claimed (based on paleoclimatic assumptions) that chickens were domesticated in Southern China in 6000 BC. However, according to a recent study, it is unclear whether those birds were the ancestors of chickens today. Instead, the origin could be the Harappan culture of the Indus Valley. Eventually, the chicken moved to the Tarim basin of central Asia. The chicken reached Europe (Romania, Turkey, Greece, Ukraine) about 3000 BC. Introduction into Western Europe came far later, about the 1st millennium BC. Phoenicians spread chickens along the Mediterranean coasts, to Iberia. Breeding increased under the Roman Empire, and was reduced in the Middle Ages. Middle East traces of chicken go back to a little earlier than 2000 BC, in Syria; chicken went southward only in the 1st millennium BC. The chicken reached Egypt for purposes of cock fighting about 1400 BC, and became widely bred only in Ptolemaic Egypt (about 300 BC). Little is known about the chicken's introduction into Africa. Three possible routes of introduction in about the early first millennium AD could have been through the Egyptian Nile Valley, the East Africa Roman-Greek or Indian trade, or from Carthage and the Berbers, across the Sahara. The earliest known remains are from Mali, Nubia, East Coast, and South Africa and date back to the middle of the first millennium AD. Domestic chicken in the Americas before Western conquest is still an ongoing discussion, but blue-egged chickens, found only in the Americas and Asia, suggest an Asian origin for early American chickens.

A lack of data from Thailand, Russia, the Indian subcontinent, Southeast Asia and Sub-Saharan Africa makes it difficult to lay out a clear map of the spread of chickens in these areas; better description and genetic analysis of local breeds threatened by extinction may also help with research into this area.

## South America

An unusual variety of chicken that has its origins in South America is the araucana, bred in southern Chile by the Mapuche people. Araucanas, some of which are tailless and some of which have tufts of feathers around their ears, lay blue-green eggs. It has long been suggested that they pre-date the arrival of European chickens brought by the Spanish and are evidence of pre-Columbian trans-Pacific contacts between Asian or Pacific Oceanic peoples, particularly the Polynesians, and South America. In 2007, an international team of researchers reported the results of analysis of chicken bones found on the Arauco Peninsula in south-central Chile. Radiocarbon dating suggested that the chickens were Pre-Columbian, and DNA analysis showed that they were related to prehistoric populations of chickens in Polynesia. These results appeared to confirm that the chickens came from Polynesia and that there were transpacific contacts between Polynesia and South America before Columbus's arrival in the Americas.

However, a later report looking at the same specimens concluded:

A published, apparently pre-Columbian, Chilean specimen and six pre-European Polynesian specimens also cluster with the same European/Indian subcontinental/Southeast Asian sequences, providing no support for a Polynesian introduction of chickens to South America. In contrast, sequences from two archaeological sites on Easter Island group with an uncommon haplogroup from Indonesia, Japan, and China and may represent a genetic signature of an early Polynesian dispersal. Modeling of the potential marine carbon contribution to the Chilean archaeological specimen casts further doubt on claims for pre-Columbian chickens, and definitive proof will require further analyses of ancient DNA sequences and radiocarbon and stable isotope data from archaeological excavations within both Chile and Polynesia.

## Farming

A former battery hen, five days after release. Note the pale comb - the comb may be a reliable indicator of health or vigor.

More than 50 billion chickens are reared annually as a source of food, for both their meat and their eggs.

The vast majority of poultry are raised in factory farms. According to the Worldwatch Institute, 74 percent of the world's poultry meat and 68 percent of eggs are produced this way. One alternative to intensive poultry farming is free range farming.

Friction between these two main methods has led to long-term issues of ethical consumerism. Opponents of intensive farming argue that it harms the environment, creates human health risks and is inhumane. Advocates of intensive farming say that their highly efficient systems save land and food resources owing to increased productivity, and that the animals are looked after in state-of-the-art environmentally controlled facilities.

In part due to the conditions on intensive poultry farms and recent recalls of large quantities of eggs, there is a growing movement for small-scale micro-flocks or 'backyard chickens'. This involves keeping small numbers of hens (usually no more than a dozen) in suburban or urban residential areas to control bugs, to utilize chicken waste as fertilizer in small gardens, and of course for the high-quality eggs and meat that are produced.

## Reared for Meat

Chickens farmed for meat are called broiler chickens. Chickens will naturally live for six or more years, but broiler chickens typically take less than six weeks to reach slaughter size. A free range or organic meat chicken will usually be slaughtered at about 14 weeks of age.

## Reared for Eggs

Chickens farmed for eggs are called egg-laying hens. In total, the UK alone consumes over 29 million eggs per day. Some hens breeds can produce over 300 eggs per year, with "the highest authenticated rate of egg laying being 371 eggs in 364 days". After 12 months of laying, the commercial hen's egg-laying ability starts to decline to the point where the flock is unviable. Hens, particularly from battery cage systems, are sometimes infirm or have lost a significant amount of their feathers, and their life expectancy has been reduced from around seven years to less than two years. In the UK and Europe, laying hens are then slaughtered and used in processed foods or sold as "soup hens". In some other countries, flocks are sometimes force moulted, rather than being slaughtered, to reinvigorate egg-laying. This involves complete withdrawal of food (and sometimes water) for 7–14 days or sufficiently long to cause a body weight loss of 25 to 35%, or up to 28 days under experimental conditions. This stimulates the hen to lose her feathers, but also reinvigorates egg-production. Some flocks may be force moulted several times. In 2003, more than 75% of all flocks were moulted in the US.

## Artificial Incubation

Incubation can successfully occur artificially in machines that provide the correct, controlled environment for the developing chick. The average incubation period for chickens is 21 days but may depend on the temperature and humidity in the incubator. Temperature regulation is the most critical factor for a successful hatch. Variations of more than 1 °C (1.8 °F) from the optimum temperature of 37.5 °C (99.5 °F) will reduce hatch rates. Humidity is also important because the rate at which eggs lose water by evaporation depends on the ambient relative humidity. Evaporation can be assessed by candling, to view the size of the air sac, or by measuring weight loss. Relative humidity should be increased to around 70% in the last three days of incubation to keep the membrane around the hatching chick from drying out after the chick cracks the shell. Lower humidity is usual in the first 18 days to ensure adequate evaporation. The position of the eggs in the incubator can also influence hatch rates. For best results, eggs should be placed with the pointed ends down and turned regularly (at least three times per day) until one to three days before hatching. If the eggs aren't turned, the embryo inside may stick to the shell and may hatch with physical defects. Adequate ventilation is necessary to provide the embryo with oxygen. Older eggs require increased ventilation.

Many commercial incubators are industrial-sized with shelves holding tens of thousands of eggs at a time, with rotation of the eggs a fully automated process. Home incubators are boxes holding from 6 to 75 eggs; they are usually electrically powered, but in the past some were heated with an oil or paraffin lamp.

## As Pets

Some people find chickens' behaviour entertaining and educational and they are sometimes kept as pets. Chickens can be tamed by hand feeding, although roosters can sometimes become aggressive and noisy. Aggression may be curbed with proper handling. Some have advised against keeping them around very young children. Certain breeds, such as silkies and many bantam varieties, are generally docile and are often recommended as good pets around children with disabilities.

## Diseases and Ailments

Chickens are susceptible to several parasites, including lice, mites, ticks, fleas, and intestinal worms, as well as other diseases. Despite the name, they are not affected by chickenpox, which is generally restricted to humans.

Some of the diseases that can affect chickens are shown below:

| Name | Common name | Cause |
|---|---|---|
| Aspergillosis | | fungi |
| Avian influenza | bird flu | virus |
| Histomoniasis | blackhead disease | protozoal parasite |
| Botulism | | toxin |
| Cage layer fatigue | | mineral deficiencies, lack of exercise |
| Campylobacteriosis | | tissue injury in the gut |
| Coccidiosis | | parasites |
| Colds | | virus |
| Crop bound | | improper feeding |
| *Dermanyssus gallinae* | red mite | parasite |
| Egg bound | | oversized egg |
| Erysipelas | | bacteria |
| Fatty liver hemorrhagic syndrome | | high-energy food |
| Fowl cholera | | bacteria |
| Fowl pox | | virus |
| Fowl typhoid | | bacteria |
| Gallid herpesvirus 1 or infectious laryngotracheitis | | virus |
| Gapeworm | *Syngamus trachea* | worms |
| Infectious bronchitis | | virus |

| | | |
|---|---|---|
| Infectious bursal disease | Gumboro | virus |
| Infectious coryza | | bacteria |
| Lymphoid leukosis | | avian leukosis virus |
| Marek's disease | | virus |
| Moniliasis | yeast infection or thrush | fungi |
| Mycoplasmas | | bacteria-like organisms |
| Newcastle disease | | virus |
| Necrotic enteritis | | bacteria |
| Omphalitis | mushy chick disease | umbilical cord stump |
| Peritonitis | | infection in abdomen from egg yolk |
| Prolapse | | |
| Psittacosis | | bacteria |
| Pullorum | salmonella | bacteria |
| Scaly leg | | parasites |
| Squamous cell carcinoma | | cancer |
| Tibial dyschondroplasia | speed growing | |
| Toxoplasmosis | | protozoal parasite |
| Ulcerative enteritis | | bacteria |
| Ulcerative pododermatitis | bumblefoot | bacteria |

## In Religion and Mythology

Vatican Persian Cock — A 1919 print of a fabric square of a Persian cock or a Persian bird design belonging to the Vatican (Holy See) in Rome dating to 600 CE. Notice the halo denoting the status of being holy in that religious schema.

Since antiquity chickens have been, and still are, a sacred animal in some cultures and deeply embedded within belief systems and religious worship. The term "Persian bird" for the cock appears to have been given by the Greeks after Persian contact "because of his great importance and his religious use among the Persians".

In Indonesia the chicken has great significance during the Hindu cremation ceremony. A chicken is considered a channel for evil spirits which may be present during the ceremony. A chicken is tethered by the leg and kept present at the ceremony for its duration to ensure that any evil spirits present go into the chicken and not the family members. The chicken is then taken home and returns to its normal life.

In ancient Greece, the chicken was not normally used for sacrifices, perhaps because it was still considered an exotic animal. Because of its valor, the cock is found as an attribute of Ares, Heracles, and Athena. The alleged last words of Socrates as he died from hemlock poisoning, as recounted by Plato, were "Crito, I owe a cock to Asclepius; will you remember to pay the debt?", signifying that death was a cure for the illness of life.

The Greeks believed that even lions were afraid of roosters. Several of Aesop's Fables reference this belief.

In the New Testament, Jesus prophesied the betrayal by Peter: "Jesus answered, 'I tell you, Peter, before the rooster crows today, you will deny three times that you know me.'" It happened, and Peter cried bitterly. This made the rooster a symbol for both vigilance and betrayal.

Earlier, Jesus compares himself to a mother hen when talking about Jerusalem: "O Jerusalem, Jerusalem, you who kill the prophets and stone those sent to you, how often I have longed to gather your children together, as a hen gathers her chicks under her wings, but you were not willing."

In the sixth century, Pope Gregory I declared the rooster the emblem of Christianity and another Papal enactment of the ninth century by Pope Nicholas I ordered the figure of the rooster to be placed on every church steeple.

In many Central European folk tales, the devil is believed to flee at the first crowing of a rooster.

In traditional Jewish practice, a kosher animal is swung around the head and then slaughtered on the afternoon before Yom Kippur, the Day of Atonement, in a ritual called kapparos; it is now common practice to cradle the bird and move it around the head. A chicken or fish is typically used because it is commonly available (and small enough to hold). The sacrifice of the animal is to receive atonement, for the animal symbolically takes on all the person's sins in kapparos. The meat is then donated to the poor. A woman brings a hen for the ceremony, while a man brings a rooster. Although not a sacrifice in the biblical sense, the death of the animal reminds the penitent sinner that his or her life is in God's hands.

The Talmud speaks of learning "courtesy toward one's mate" from the rooster. This might refer to the fact that when a rooster finds something good to eat, he calls his hens to eat first. The Talmud likewise provides us with the statement "Had the Torah not been given to us, we would have learned modesty from cats, honest toil from ants, chastity from doves and gallantry from cocks," which may be further understood as to that of the gallantry of cocks being taken in the context of a religious instilling vessel of "a girt one of the loins" (Young's Literal Translation) that which is "stately in his stride" and "move with stately bearing" in the Book of Proverbs 30:29-31 as referenced by Michael V. Fox in his *Proverbs* 10-31 where Sa'adiah ben Yosef Gaon (Saadia Gaon) identifies the definitive trait of "A cock girded about the loins" in Proverbs 30:31 (Douay–Rheims Bible) as "the honesty of their behavior and their success", identifying a spiritual purpose of a religious vessel within that religious instilling schema of purpose and use.

The chicken is one of the Zodiac symbols of the Chinese calendar. In Chinese folk religion, a cooked chicken as a religious offering is usually limited to ancestor veneration and worship of village deities. Vegetarian deities such as the Buddha are not recipients of such offerings. Under some observations, an offering of chicken is presented with "serious" prayer (while roasted pork is offered during a joyous celebration). In Confucian Chinese weddings, a chicken can be used as a substitute for one who is seriously ill or not available (e.g., sudden death) to attend the ceremony. A red silk scarf is placed on the chicken's head and a close relative of the absent bride/groom holds the chicken so the ceremony may proceed. However, this practice is rare today.

A cockatrice was supposed to have been born from an egg laid by a rooster, as well as killed by a rooster's call.

## In History

The red junglefowl

An early domestication of chickens in Southeast Asia is probable, since the word for domestic chicken (**manuk*) is part of the reconstructed Proto-Austronesian language. Chickens, together with dogs and pigs, were the domestic animals of the Lapita culture, the first Neolithic culture of Oceania.

The first pictures of chickens in Europe are found on Corinthian pottery of the 7th century BC. The poet Cratinus (mid-5th century BC, according to the later Greek author Athenaeus) calls the chicken "the Persian alarm". In Aristophanes's comedy *The Birds* (414 BC) a chicken is called "the Median bird", which points to an introduction from the East. Pictures of chickens are found on Greek red figure and black-figure pottery.

In ancient Greece, chickens were still rare and were a rather prestigious food for symposia. Delos seems to have been a center of chicken breeding (Columella, *De Re Rustica* 8.3.4).

The Romans used chickens for oracles, both when flying ("ex avibus", Augury) and when feeding ("auspicium ex tripudiis", Alectryomancy). The hen ("gallina") gave a favourable omen ("auspicium ratum"), when appearing from the left (Cic.,de Div. ii.26), like the crow and the owl.

For the oracle "ex tripudiis" according to Cicero (Cic. de Div. ii.34), any bird could be used in auspice, and shows at one point that any bird could perform the *tripudium* but normally only chickens ("pulli") were consulted. The chickens were cared for by the pullarius, who opened their cage and fed them pulses or a special kind of soft cake when an augury was needed. If the chickens stayed in their cage, made noises ("occinerent"), beat their wings or flew away, the omen was bad; if they ate greedily, the omen was good.

In 249 BC, the Roman general Publius Claudius Pulcher had his "sacred chickens" " thrown overboard when they refused to feed before the battle of Drepana, saying "If they won't eat, perhaps they will drink." He promptly lost the battle against the Carthaginians and 93 Roman ships were sunk. Back in Rome, he was tried for impiety and heavily fined.

In 162 BC, the Lex Faunia forbade fattening hens to conserve grain rations. To get around this, the Romans castrated roosters (capon), which resulted in a doubling of size despite the law that was passed in Rome that forbade the consumption of fattened chickens. It was renewed a number of times, but does not seem to have been successful. Fattening chickens with bread soaked in milk was thought to give especially delicious results. The Roman gourmet Apicius offers 17 recipes for chicken, mainly boiled chicken with a sauce. All parts of the animal are used: the recipes include the stomach, liver, testicles and even the pygostyle (the fatty "tail" of the chicken where the tail feathers attach).

The Roman author Columella gives advice on chicken breeding in his eighth book of his treatise, *De Re Rustica* (On Agriculture). He identified Tanagrian, Rhodic, Chalkidic and Median (commonly misidentified as Melian) breeds, which have an impressive appearance, a quarrelsome nature and were used for cockfighting by the Greeks (*De Re Rustica* 8.3.4). For farming, native (Roman) chickens are to be preferred, or a cross between native hens and Greek cocks (*De Re Rustica* 8.2.13). Dwarf chickens are nice to watch because of their size but have no other advantages.

According to Columella (*De Re Rustica* 8.2.7), the ideal flock consists of 200 birds, which can be supervised by one person if someone is watching for stray animals. White chickens should be avoided as they are not very fertile and are easily caught by eagles or goshawks. One cock should be kept for five hens. In the case of Rhodian and Median cocks that are very heavy and therefore not much inclined to sex, only three hens are kept per cock. The hens of heavy fowls are not much inclined to brood; therefore their eggs are best hatched by normal hens. A hen can hatch no more than 15-23 eggs, depending on the time of year, and supervise no more than 30 hatchlings. Eggs that are long and pointed give more male, rounded eggs mainly female hatchlings (*De Re Rustica* 8.5.11).

Columella also states that chicken coops should face southeast and lie adjacent to the kitchen, as smoke is beneficial for the animals and "poultry never thrive so well as in warmth and smoke" (*De Re Rustica* 8.3.1). Coops should consist of three rooms and possess a hearth. Dry dust or ash should be provided for dust-baths.

According to Columella (*De Re Rustica* 8.4.1), chickens should be fed on barley groats, small chick-peas, millet and wheat bran, if they are cheap. Wheat itself should be avoided as it is harmful to the birds. Boiled ryegrass (*Lolium* sp.) and the leaves and seeds of alfalfa (*Medicago sativa* L.) can be used as well. Grape marc can be used, but only when the hens stop laying eggs, that is, about the middle of November; otherwise eggs are small and few. When feeding grape marc, it should be supplemented with some bran. Hens start to lay eggs after the winter solstice, in warm places around the first of January, in colder areas in the middle of February. Parboiled barley increases their fertility; this should be mixed with alfalfa leaves and seeds, or vetches or millet if alfalfa is not at hand. Free-ranging chickens should receive two cups of barley daily.

Columella advises farmers to slaughter hens that are older than three years, because they no longer produce sufficient eggs.

According to Aldrovandi: Capons were produced by burning "the hind part of the bowels, or loins or spurs" with a hot iron. The wound was treated with potter's chalk.

For the use of poultry and eggs in the kitchens of ancient Rome see Roman eating and drinking.

Chickens were spread by Polynesian seafarers and reached Easter Island in the 12th century AD, where they were the only domestic animal, with the possible exception of the Polynesian rat (*Rattus exulans*). They were housed in extremely solid chicken coops built from stone, which was first reported as such to Linton Palmer in 1868, who also "expressed his doubts about this".

## As Food

The meat of the chicken, also called "chicken", is a type of poultry meat. Because of its relatively low cost, chicken is one of the most used meats in the world. Nearly all parts of the bird can be used for food, and the meat can be cooked in many different ways. Popular chicken dishes include roasted chicken, fried chicken, chicken soup, Buffalo wings, tandoori chicken, butter chicken, and chicken rice. Chicken is also a staple of many fast food restaurants.

### Eggs

In 2000, there were 50.4 million tons of eggs produced in the world (*Executive guide to world poultry trends,* 2001) and an estimated 53.4 million tons of table eggs were produced during 2002. In 2009, an estimated 62.1 million metric tons of eggs were produced worldwide from a total laying flock of approximately 6.4 billion hens.

Chicken eggs are widely used in many types of dishes, both sweet and savory, including many baked goods. Eggs can be scrambled, fried, hard-boiled, soft-boiled, pickled, and poached. The albumen, or egg white, contains protein but little or no fat, and can be used in cooking separately from the yolk. Egg whites may be aerated or whipped to a light, fluffy consistency and are often used in desserts such as meringues and mousse. Ground egg shells are sometimes used as a food additive to deliver calcium. Hens do not need a male to produce eggs, only to fertilize them. A flock containing only females will still produce eggs; however, the eggs will all be infertile.

## Andalusian Chicken

Blue Andalusian cock

The Andalusian or Blue Andalusian, Spanish: Andaluza Azul, is a breed of domestic chicken indigenous to the autonomous community of Andalusia in south-west Spain. It is distributed through much of the countryside of Córdoba and Seville, and is concentrated particularly in the area of Utrera, which is considered the heartland of the breed. In 2009 the population was estimated at 10,000 birds.

Blue Andalusian hen

A very different type of Andalusian, more intensely blue and with blue-laced plumage, was created in England from birds imported from Andalusia through selective breeding and cross-breeding with birds of other breeds.

## History

There is little information on the early history of the Andalusian. Blue chickens from Andalusia were imported to England no later than 1851. The creation of the "international" type of Andalusian, with blue-laced plumage, is attributed to the English, whether in Andalusia or in Britain. Two breeders in particular are thought to have started this process, which took many years: one named Coles, from Fareham, Hampshire, and a certain John Taylor of Shepherd's Bush, in west London. Andalusians were shown at Baker Street, London, in January 1853; they were not included in the original *Standard of Excellence* in 1865.

Andalusians reached the United States in about 1850–55, and were included in the first edition of the *Standard of Perfection* of the American Poultry Association in 1874. The breed arrived in South America in 1870, and was first shown in Germany in the same year. A bantam was created in the 1880s.

## Characteristics

The slate-blue plumage of the Andalusian is caused by a dilution gene, which, in combination with the E gene for black plumage, produces partial dilution of the melanin which gives the black colour. Not all Andalusians are blue: birds with two copies of the gene have near-total dilution, and are off-white; birds with no copies have no dilution, and are black; those with one copy have partial dilution, and are blue. Blue birds occur, in Mendelian proportion, twice as often as each of the other colours. All are present in the population.

The earlobes of the Andalusian are smooth, white, and almond-shaped; the crest is single and of medium size, with five well-defined points. The skin is white and the legs and feet are black.

### Use

Andaluza Azul hens lay about 165 white eggs per year; eggs weigh 70–80 g. Blue-bred white hens lay the largest eggs.

## Bresse Gauloise

The Bresse Gauloise is a French breed of domestic chicken. It originates in the historic region and former province of Bresse, in the regions of Rhône-Alpes, Bourgogne and Franche-Comté, in eastern France. Because of legal restrictions on the use of the name, only white chickens raised within that area may be called "Bresse"; outside it, they are given the name "Gauloise"; the breed name combines both. Four colours are recognised for the Bresse Gauloise, three of them linked to areas within Bresse: the Bresse de Bourg is "grey" (silver-pencilled); the Bresse de Bény is white; the Bresse de Louhans is black; a blue variety has recently been created. White Bresse de Bény chickens and capons raised in the area of Bresse have *appellation d'origine contrôlée* status and are marketed as poulet de Bresse; they are regarded as a premium product and command higher prices than other chickens.

### History

The first documentation of the chickens of Bresse reportedly dates from 12 November 1591, when the citizens of Bourg presented two dozen birds to Joachim de Rye, Marquis de Treffort. In the early nineteenth century, the lawyer, politician, epicure and gastronome Jean Anthelme Brillat-Savarin (1755–1826), who was born at Belley in the Ain, is supposed to have described the Bresse chicken as "the queen of poultry, the poultry of kings". Like the La Flèche, which was raised and fattened in a similar fashion, the Bresse chicken had high standing in the market. Nevertheless, by about 1900 the breed had virtually disappeared. Its recovery was due to fancy breeders, who selectively bred a sufficient number of white chickens for the breed to become stable. A new breed standard was drawn up in 1904. The Bresse name, used for both chicken products and for the dinde de Bresse, the turkey of the area, received legal protection on 22 December 1936; this became an *appellation d'origine contrôlée* (AOC) in 1957.

### Characteristics

### Use

Although the Bresse Gauloise is principally famous as a meat breed, it is a good layer of large white eggs, which in the black Louhans variety may weigh 70 g. Hens are usually non-sitting.

# Delaware Chicken

A Delaware hen

The Delaware is a breed of chicken originating in the U.S. state of Delaware. It was once of relative importance to the U.S. chicken industry, but today is critically endangered. It is primarily suited to meat production but also lays reasonably well. It has plumage of a unique pattern, and is accepted into poultry standards for showing.

## Characteristics

With males weighing 8.5 pounds (3.9 kilos) and hens 6.5 pounds (3 kilos), the Delaware is a medium sized breed. They have rather large, bright red colored single combs and wattles. Delawares appear in a single color type: a white body and breast, with light black barring on the ends of the hackle, wings and tails. It is similar to the Columbian color seen in some breeds, but has barring in the dark portions, rather than uniform black. Also of note is that all feathers have a white quill and shaft, which, combined with yellow skin, makes for a cleaner appearing carcass. Like most standard breeds of chicken, the Delaware has a miniaturized bantam version; however, these are rarely seen.

Delawares are hardy birds that mature quickly. Hens are good layers of large to jumbo brown eggs and will go broody. Unlike the most common commercial meat birds in use today, the Delaware does well in free range operations. In temperament, it is a calm, but not a typically friendly bird although in some cases can be.

## Approximate Weight

| | | |
|---|---|---|
| Rooster | 3.9 kg | 8.5 lbs |
| Hen | 2.9 kg | 6.5 lbs |
| BANTAM | | |
| Rooster | 32 oz | |
| Hen | 28 oz | |

## History

In the early 20th century, crosses of Barred Plymouth Rock roosters on New Hampshire hens was a common choice for producing broilers. Occasionally, this mating produces sports with light coloration. By breeding these white (genetically silver) sports intentionally, George Ellis of Delaware created the breed in 1940. He first chose to call them *Indian Rivers*, but later the name was switched to match its state of origin. At the time, the Delmarva Peninsula, where the breed was created, supplied chicken to the entirety of the East Coast of the United States through companies such as Perdue Farms. The Delaware rapidly became the premiere broiler fowl in use in the region, thus affecting the industry at large. In 1952, it was recognized for exhibition by acceptance into the American Poultry Association's Standard of Perfection.

Beginning in the mid-1950s, commercial farms began to use the White Cornish-Rock cross that would come to dominate the chicken industry into the next century. The speedy adoption of the Cornish-Rock saw the decline of the Delaware, though it persisted in some areas into the 1960s. In the 21st century, the Delaware is considered a critically endangered breed by organizations such as the American Livestock Breeds Conservancy. It is also included in Slow Food USA's Ark of Taste, a list of heritage foods.

# Asil Chicken

The Asil or Aseel is a breed of chicken originating from South Punjab / Sindh area of Pakistan and India.Also In Iran Called Lari most famous Rooster in Iran. fowl are found throughout Southeast Asia, such as Shamo and Thai Game. The breed is generally unstandardised in South Asia and India, but popularity has increased in the western world in recent times with the breed standardised in the British, Australian and American standards.

A sawa aseel bred by Azhar Khan

A Reza Asil rooster and two hens

long tail parrot beak asil

## Description and History

Asils were first used fighting cocks. Aseel is noted for its pugnacity. The chicks often fight when they are just a few weeks old and mature roosters will fight to the death. Hens can also be very aggressive towards each other.

Towards humans Asil are generally very tame and trusting. There are anecdotes where they have come to their keepers for other things than food, for example to get the keeper to open the door to the coop so they can get to roost.

The hens are not good layers, but are excellent sitters. Laying depends on the Asil variety, the small Asil are known to be very poor layers, sometimes laying just 6 eggs a year, whereas larger Asil can lay around 40 eggs a year.

In the U.S., the breed is on the "Watchlist" (2012), by the Livestock Conservancy. The Aseel breed is found in almost all states of India, but abundant in Andhra Pradesh.

## Varieties

Portrait of Kulang Asil rooster head

There are many varieties of Asil, some are standardized for shows such as the Reza Asil in the UK, some are simply named after the area where they are bred such as the Mianwali Asil from Pakistan or the colour, red/wheaten Asil are generally known as "Sonatol".

There are also hen-feathered Asil knows as "Madaroo" these are found in various colours, but the cocks come with feathers in hen colour, don't have sickle feathers in the tails and miss the large hanging feathers on the saddle. This variety is very rare.

Asil with feather beards under their beaks known as "muffed" and with tufts on the top of their heads known as "tasseled" are also seen, but are very rare especially outside India/Pakistan.

Bhaingam Asil variety have a large single comb but confirm to all the other Asil standards.

Broadly speaking, Asil in Europe are categorized and shown under these three types:

## Reza Asil

Height: Up to 50 cm tall. Weight: Maximum weight for the hens is 1.8 kg, max weight for the cocks is 2.7 to 3 kg.

This type is standardized by the Asian Hardfeather Society in the UK and is seen at shows throughout the UK, but is quite rare.

This group of Asil reached worldwide popularity due to books and articles written by gamefowl experts such as Herbert Atkinson, Siran and Paul Deraniyagala from Sri

Lanka and Carlos Finsterbusch from Chile. The Reza Asil family according to the old (Western) gamefowl literature is subdivided into following strains: (Amir) Ghan (Dark-Red), Sonatol(Light-Red), (Siyah) Rampur(Black), Kalkatiya (Kaptan)(Speckled-Reds) and Jawa(Duckwing). All these strains are identified by their specific color, these colors do not necessarily correspond with the area where the birds come from.

In colonial times other colors such as whites, spangles, golden etc. were regarded as inferior. At present day the "classic" strains and names given mentioned by Atkinson are more or less forgotten. The native people in India, Pakistan, Bangladesh and Sri Lanka only know the Reza-type Asil by their local names.

## Reza Asil also Available in Maharashtra

Reza Asil available in Satona (Kh) 431502 Tq. Partur Dis. Jalna, Selgaon Tq. Partur, Malegaon Tq. Dhule Dist. Nasik

## Kulang Asil

Height: Up to 75 cm tall. Weight: 5 to 7 kg.

The large Asil are divided into sub-varieties : North Indian, South Indian and Madras type. The North and South Indian varieties don't differ much. Only type of comb, shape of the beak and body shape are different. For example: Northern type = slender, Southern type = heavier build. The Madras Asil however is significantly different. They have a lower station, are heavier build and stronger boned. These birds often come in a bluish colour. This variety is found in the deep south of India, the Tamil Nadu state.

## Sindhi Aseel

A 5 months old Sindhi Kaura Aseel

Sindhi Aseel or Sindhi Asil is a breed of chicken and as from the name, originates from Sindh (one of 4 major provinces of Pakistan). These cocks, or fighting cocks, are noted being tall, heavy and good at fighting, so they are bred mainly for cockpit. These Aseels are characterised by a muscular but compact body, broad shoulders, wings carried against the body, short and hard feathers, drooping tails, a large curved beak similar to that of an eagle, pea comb and no wattles.

## Mianwali

This breed is mainly found in Mianwali district of Pakistan. However, since its arrival, this breed has risen to popularity in Pakistan, currently the primary game breed used in the pits also preferred by gamblers. It is smaller compared to Sindhi aseels weighing between 1.5 and 3.5 kg depending on the preference of breeders. It is much faster and a better head hitter usually comes in small to medium height. A good Mianwali aseel should kill its opponent within a few minutes. They have been known to kill bigger roosters because of their speed and accuracy. They come in various colors such as Java(duckwing), Lakha (reddish), black and various others depending on the combination used in breeding. Very hard and a brave fighter with attitude to inspire, excellent in naked heels and metal spurs. There are many sub breeds of this breed owing to the combination used in breeding. A good tested Mianwali rooster would usually have offspring of a similar quality. Typical description would be small curved beak, strong joints, pearl/white/yellow eye color, short crow, small comb and do not have heavy body structure. May look smaller than other breeds but is excellent spurer.

## Java Aseel

Java aseel is one of the most used and best fighter cock in Pakistan/India. It is mainly found in Punjab region, its origin was from Mianwali district, Jawa Aseel is recognized by his black and yellow colour.

java breed asil

It is also found in red and black colours and orangish most of the times. It is usually mixed with desi cock breeds for superior fighting skills. It's weak point is its long legs and eta more hits in fighting.. it is also found in South India in Kolar district of Karnataka belts Height of Java Asil is 50 cm and Weight of Java is 7 to 8 kg.

### Amroha

This is a rare breed of Aseel used in Pakistan and India. Very few of these roosters exist in their pure form. They are known to be small to medium like Mianwali. It is also known that they are champion of naked heel fighting. In simple, it is a fantasy of most aseel breeders in Philippines. they are very strong .mostly they found in two colors.

### Bantam Asil

Weight: Up to 0.75 kg.

Bantam Asil have been created at the end of the 19th century by an English breeder named William Flamank Entwisle. The breed got very popular after its creation but after a couple of decades interest in this variety slowly died out. Until the beginning of the 1980s nothing was heard about these little Asil. A Belgian breeder named Willy Coppens created them again using Shamo (chicken), Indian Game and Reza Asil. The breed was also introduced again in the Netherlands and United Kingdom. At present day Bantam Asil are quite popular and they are bred in various colors.

### Lasani Aseel Breed

Pakistani Aseel breeders have developed different varieties of aseel breeds. But one of the rarest breed is what they call Lassani Aseel breed. They are medium size birds with powerful strikes to earn them the title of neck breakers. This breed has a typical parrot like beak and small neck size. Unlike mianwali they have a particular fight style. They don't like jumping over the opponent rather they like to come closer to the opponent touching him with their front body and than attacking on the neck. So in a typical mianwali versus Lassani fight you will see mianwali birds dominating initially but as soon as they are tired of jumping around and land on the ground somewhat tired lasani breed cock will finish the game with a couple of hits in the neck. Amroha and lassani are closely related however lassani breeds have smaller necks and often closed eyes only few families of Nawab have got this breed in its pure form. And the notion that they are bengam type aseels is mistaken.

## Broiler

Broiler chickens (*Gallus gallus domesticus*), or broilers, are a gallinaceous domesticated fowl, bred and raised specifically for meat production. They are a hybrid of the egg-laying chicken, both being a subspecies of the red junglefowl (*Gallus gallus*). Typical broilers have white feathers and yellowish skin. Most commercial broilers reach slaughter-weight at between five and seven weeks of age, although slower growing breeds reach slaughter-weight at approximately 14 weeks of age. Because the meat broilers are this young at slaughter, their behaviour and physiology are that of an immature bird. Due to artificial selection for rapid early growth and the husbandry used to sustain this, broilers are susceptible to several welfare concerns, particularly skeletal malformation and dysfunction, skin and eye lesions, and congestive heart conditions. The breeding stock (broiler-breeders) grow to maturity and beyond but also have welfare issues related to frustration of a high feeding motivation and beak trimming. Broilers are usually grown as mixed-sex flocks in large sheds under intensive conditions, but some breeds can be grown as free-range flocks. Chickens are one of the most common and widespread domestic animals.

## Origins and Domestication

A male red junglefowl (*Gallus gallus*) - the ancestral precursor of the modern broiler.

The domestic chicken is descended primarily from the red junglefowl (*Gallus gallus*) and is classified as a sub-species (*Gallus gallus domesticus*) of that species. As such, it can and does freely interbreed with populations of red jungle fowl.

The traditional poultry farming view is stated in *Encyclopædia Britannica* (2007): "Humans first domesticated chickens of Indian origin for the purpose of cockfighting in Asia, Africa, and Europe. Very little formal attention was given to egg or meat production... ".

Genetic studies point to multiple maternal origins, with the clade found in the Americas, Europe, Middle East, and Africa, originating from the Indian subcontinent, where a large number of unique haplotypes occur. It is postulated that the jungle fowl, known as the "bamboo fowl" in many Southeast Asian languages, is a pheasant well adapted to take advantage of the large amounts of fruits that are produced during the end of the 50 year bamboo seeding cycle to boost its own reproduction. It has been suggested that in domesticating the chicken, humans took advantage of this prolific reproduction of the jungle fowl when exposed to large amount of food. Similarly, it is speculated that the flexibility and adaptability inherited from the chicken's red junglefowl ancestor allows them to cope with the "unnatural and intense conditions" of modern production.

Recent genetic analysis has revealed that at least the gene for yellow skin was incorporated into domestic birds through hybridization with the grey junglefowl (*G. sonneratii*).

## Modern Breeding

Before the development of modern commercial meat breeds (chickens, etc.), broilers were mostly young male chickens culled from farm flocks. Pedigree breeding began around 1916. Magazines for the poultry industry existed at this time. A hybrid variety of chicken was produced from a cross of a male of a naturally double-breasted Cornish strain and a female of a tall, large-boned strain of white Plymouth Rocks. This first attempt at a hybrid meat breed was introduced in the 1930s and became dominant in the 1960s. The original hybrid was plagued by problems of low fertility, slow growth and disease susceptibility.

Modern broilers have become very different from the Cornish/Rock hybrid. As an example, Donald Shaver (originally a breeder of egg-layers) began gathering breeding stock for a broiler program in 1950. Besides the breeds normally favoured, Cornish Game, Plymouth Rock, New Hampshire, Langshans, Jersey Black Giant and Brahmas were included. A white feathered female line was purchased from Cobb. A full scale breeding program was commenced in 1958, with commercial shipments in Canada and the US in 1959 and in Europe in 1963.

As a second example, colour sexing broilers was proposed by Shaver in 1973. The genetics were based on the company's breeding plan for egg-layers which had been developed in the mid-1960s. A difficulty facing the breeders of the colour-sexed broiler is that the chicken must be white-feathered by slaughter age. After 12 years, accurate colour sexing without compromising economic traits was achieved.

## General Biology

A one-day-old chick.

Modern commercial broilers, for example, Cornish crosses and Cornish-Rocks, are artificially selected and bred for large-scale, efficient meat production and although they are the same species, grow much faster than egg laying hens or traditional dual purpose breeds. They are noted for having very fast growth rates, a high feed conversion ratio, and low levels of activity. Modern commercial broilers are bred to reach a slaughter-weight of about 2 kg in only 35 to 49 days. Slow growing free-range and organic strains have been developed which reach slaughter-weight at 12 to 16 weeks of age. As a consequence, the behaviour and physiology of broilers reared for meat are those of immature birds, rather than adults.

Typical broilers have white feathers and yellowish skin. Recent genetic analysis has revealed that the gene for yellow skin was incorporated into domestic birds through hybridization with the grey junglefowl (*G. sonneratii*). Modern crosses are also favorable for meat production because they lack the typical "hair" which many breeds have that necessitates singeing after plucking. Both male and female broilers are reared for their meat.

## Behaviour

Because broiler chickens are the same species as egg laying hens, their behavioural repertoires are initially similar, and also similar to those of other gallinaceous birds. However, broiler behaviour is modified by the environment and alters as the broilers' age and bodyweight rapidly increase. For example, the activity of broilers reared outdoors is initially greater than broilers reared indoors, but from six weeks of age, decreases to comparable levels in all groups. The same study shows that in the outdoors group, surprisingly little use is made of the extra space and facilities such as perches – it was proposed that the main reason for this was leg weakness as 80 per cent of the birds had a detectable gait abnormality at seven weeks of age. There is no evidence of reduced motivation to extend the behavioural repertoire, as, for example, ground pecking remained at significantly higher levels in the outdoor groups because this behaviour could also be performed from a lying posture rather than standing.

Broiler breeders, i.e. males and females reared to fertilise and lay the eggs of the offspring reared for food, perform similar mating behaviour to other chicken types. They exhibit male-male aggression, male-hen aggression, hen-hen aggression, male waltzes, hen crouches, attempted hen mounts, completed hen mounts, attempted hen matings, and completed hen matings. These behaviours are seen less often and may not be exhibited as vigorously as observed in other chicken types. Examining the frequency of all sexual behaviour shows a large decrease with age, suggestive of a decline in libido. The decline in libido is not enough to account for reduced fertility in heavy cocks at 58 weeks and is probably a consequence of the large bulk or the conformation of the males at this age interfering in some way with the transfer of semen during copulations which otherwise look normal.

## Physiology

Nociceptors which respond to noxious stimulation have been identified and physiologically characterised in many different part of the body of the chicken including the beak, mouth, nose, joint capsule and scaly skin. Stimulation of these nociceptors produces cardiovascular and behavioural changes consistent with those seen in mammals and are indicative of pain perception. Physiological and behavioural experiments have identified the problem of acute pain following beak trimming in chicks, shackling, and feather pecking and environmental pollution.

## Feeding and Feed Conversion

Chickens are omnivores and modern broilers are given access to a special diet of high protein feed, usually delivered via an automated feeding system. This is combined with artificial lighting conditions to stimulate eating and growth and thus the desired body weight.

In the U.S. in 2011, the average feed conversion ratio of a broiler was 1.91 pounds of feed per pound of liveweight. In 1925 the figure was 4.70.

Canada has a typical FCR of 1.72.

New Zealand commercial broiler farms have recorded the world's best broiler chicken FCR, consistently at 1.38 or lower.

## Welfare Issues

## Meat Birds

One-day old chicks arriving to be unpacked and placed in shed.

Young birds being reared in a closed broiler house.

Artificial selection has led to a great increase in the speed with which broilers develop and reach slaughter-weight. The time required to reach 1.5 kg live-weight decreased from 120 days to 30 days between 1925 and 2005. Selection for fast early growth-rate, and feeding and management procedures to support such growth, have led to various welfare problems in modern broiler strains. Welfare of broilers is of particular concern given the large number of individuals that are produced; for example, the U.S. in 2011 produced approximately 9 billion broiler chickens.

## Cardiovascular Dysfunction

Selection and husbandry for very fast growth means there is a genetically induced mismatch between the energy-supplying organs of the broiler and its energy-consuming organs. Rapid growth can lead to metabolic disorders such as sudden death syndrome (SDS) and ascites.

SDS is an acute heart failure disease that affects mainly male fast-growing broilers which appear to be in good condition. Affected birds suddenly start to flap their wings, lose their balance, sometimes cry out and then fall on their backs or sides and die, usually all within a minute. In 1993, U.K. broiler producers reported an incidence of 0.8%. In 2000, SDS has a death rate of 0.1% to 3% in Europe.

Ascites is characterised by hypertrophy and dilatation of the heart, changes in liver function, pulmonary insufficiency, hypoxaemia and accumulation of large amounts of fluid in the abdominal cavity. Ascites develops gradually and the birds suffer for an extended period before they die. In the UK, up to 19 million broilers die in their sheds from heart failure each year.

## Skeletal Dysfunction

Breeding for increased breast muscle means that the broilers' centre of gravity has moved forward and their breasts are broader compared with their ancestors, which affects the way they walk and puts additional stresses on their hips and legs. There is a high frequency of skeletal problems in broilers, mainly in the locomotory system, including varus and valgus deformities, osteodystrophy, dyschondroplasia and femoral head necrosis. These leg abnormalities impair the locomotor abilities of the birds, and lame birds spend more time lying and sleeping. The behavioural activities of broilers decrease rapidly from 14 days of age onwards. Reduced locomotion also decreases ossification of the bones and results in skeletal abnormalities; these are reduced when broilers have been exercised under experimental conditions.

Most broilers find walking painful, as indicated by studies using analgesic and anti-inflammatory drugs. In one experiment, healthy birds took 11 seconds to negotiate an obstacle course, whereas lame birds took 34 seconds. After the birds had been treated with carprofen, there was no effect on the speed of the healthy birds, however, the lame birds now took only 18 seconds to negotiate the course, indicating that the pain of lameness is relieved by the drug. In self-selection experiments, lame birds select more drugged feed than non-lame birds leading to the suggestion that leg problems in broilers are painful.

Several research groups have developed "gait scores" (GS) to objectively rank the walking ability and lameness of broilers. In one example of these scales, GS=0 indicates normal walking ability, GS=3 indicates an obvious gait abnormality which affects the bird's ability to move about and GS=5 indicates a bird that cannot walk at all. GS=5 birds tried to use their wings to help them walking, or crawled along on their shanks. In one study, almost 26% of the birds examined were rated as GS=3 or above and can therefore be considered to have suffered from painful lameness.

Compassion in World Farming wrote on the incidence of leg problems in broilers:

> "...there is evidence that, far from improving, leg problems may have deteriorated further during the 1990s. Large and representative surveys of commercial broiler flocks in Denmark (1999) and Sweden (2002) found that in Denmark, 75% of the chickens had some walking abnormality and 30.1% were very lame (gait score greater than 2). In Sweden, over 72% of the chickens had some walking abnormality and around20% were very lame. 36.9% of the chickens surveyed in Denmark and around half (46.4% and 52.6%, depending on strain) of the chickens surveyed in Sweden had leg deformities (varus/valgus). 57% of the chickens surveyed in Denmark and around half of the chickens surveyed in Sweden showed some evidence of tibial dychondroplasia (Sanotra, Berg and Lund, 2003)."

## Integument Lesions

Sitting and lying behaviours in fast growing strains increase with age from 75% in the first seven days to 90% at 35 days of age. This increased inactivity is linked with an increase in dermatitis caused by a greater amount of time in contact with ammonia in the litter. This contact dermatitis is characterised by hyperkeratosis and necrosis of the epidermis at the affected sites; it can take forms such as hock burns, breast blisters and foot pad lesions.

## Stocking Density

Broilers in a rearing shed indicating the high stocking densities used.

Broilers are usually kept at high stocking densities which vary considerably between countries. Typical stocking densities in Europe range between about 22 to 42 $kg/m^2$ or between about 11 to 25 birds per square metre. There is a reduction of feed intake and reduced growth rate when stocking density exceeds approximately 30 $kg/m^2$ under deep litter conditions. The reduced growth rate is likely due to a reduced capacity to lose heat generated by metabolism. Higher stocking densities are associated with increased dermatitis including food pad lesions, breast blisters and soiled plumage. In a large-scale experiment with commercial farms, it was shown that the management conditions (litter quality, temperature and humidity) were more important than stocking density.

## Ocular Dysfunction

In attempts to improve or maintain fast growth, broilers are kept under a range of lighting conditions. These include continuous light (fluorescent and incandescent), continuous darkness, or under dim light; chickens kept under these light conditions develop eye abnormalities such as macrophthalmos, avian glaucoma, ocular enlargement and shallow anterior chambers.

## Ammonia

The litter in broiler pens can become highly polluted from the nitrogenous feces of the birds and produce ammonia. Ammonia has been shown to cause increased susceptibility to disease and other health-related problems such as Newcastle disease, airsaculitis and keratoconjunctivitis. The respiratory epithelium in birds is damaged by ammonia concentrations in the air exceeding 75 parts per million (ppm). Ammonia concentrations at 25 to 50 ppm induce eye lesions in broiler chicks after seven days of exposure.

## Catching and Transport

Once the broilers have reached the target live-weight, they are caught, usually by hand,

and packed live into crates for transport to the slaughterhouse. They are usually deprived of food and water for several hours before catching until slaughter. The process of catching, loading, transport and unloading causes serious stress, injury and even death to a large number of broilers.

The number of broilers that died in the EU in 2005 during the process of catching, packing and transport was estimated to be as high as 18 to 35 million. In the UK, of broilers that were found to be 'dead on arrival' at the slaughterhouse in 2005, it was estimated that up to 40% may have died from thermal stress or suffocation due to crowding on the transporter.

Slaughter is done by hanging the birds fully conscious by their feet upside-down in shackles on a moving chain, stunning them by automatically immersing them in an electrified water bath and exsanguination by cutting their throats.

Some research indicates that chickens might be more intelligent than previously supposed, which "raises questions about how they are treated". A possible 10-year life span has been shortened to six weeks for broilers.

## Mortality Rates

According to historic records, broiler mortality rates are decreasing. In 1925, the mortality rate in the U.S. was 18% compared to 3.8% in 2011.

One indication of the effect of broilers' rapid growth rate on welfare is a comparison of the usual mortality rate for standard broiler chickens (1% per week) with that for slower-growing broiler chickens (0.25% per week) and with young laying hens (0.14% per week); the mortality rate of the fast-growing broilers is seven times the rate of laying hens (the same subspecies) of the same age.

## Parent Birds

Meat broilers are usually slaughtered at approximately 35 to 49 days of age, well before they become sexually reproductive at 5 to 6 months of age. However, the bird's parents, often called "broiler-breeders", must live to maturity and beyond so they can be used for breeding. As a consequence, they have additional welfare concerns.

Meat broilers have been artificially selected for an extremely high feeding motivation, but are *not* usually feed-restricted, as this would delay the time taken for them to reach slaughter-weight. Broiler-breeders have the same highly increased feeding motivation, but *must* be feed-restricted to prevent them becoming overweight with all its concomitant life-threatening problems. An experiment on broilers' food intake found that 20% of birds allowed to eat as much as they wanted either died or had to be killed because of severe illness between 11 and 20 weeks of age - either they became so lame they could not stand or they developed cardiovascular problems.

Broiler breeders fed on commercial rations eat only a quarter to a half as much as they would with free access to food. They are highly motivated to eat at all times, presumably leading to chronic frustration of feeding.

Because broiler breeders live to adulthood, they might show feather pecking or other injurious pecking behaviour. To avoid this, they might be beak trimmed which can lead to acute or chronic pain.

## World Production and Consumption

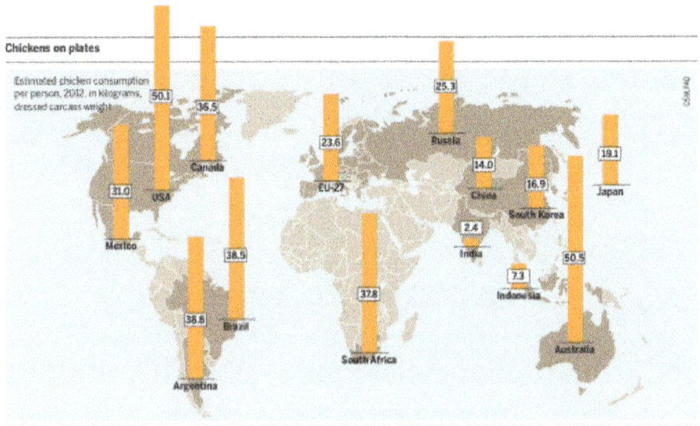


Consumption of broilers is surpassing that of beef in industrialized countries. Demand in Asia is rising.

A report in 2005 stated that around 5.9 billion broiler chickens for eating were produced yearly in the European Union. Mass production of chicken meat is a global industry and at that time, only two or three breeding companies supplied around 90% of the world's breeder-broilers. The total number of meat chickens produced in the world was nearly 47 billion in 2004; of these, approximately 19% were produced in the US, 15% in China, 13% in the EU25 and 11% in Brazil.

Worldwide, 86.6 million tonnes of broiler meat will be produced in 2014.

## Broiler Industry

The commercial production of broiler chickens for meat consumption is a highly industrialized process. There are two major sectors: (1) rearing birds intended for consumption and (2) rearing parent stock for breeding the meat birds.

# Domestic Duck

Domesticated ducks are ducks that are raised for meat, eggs and down. Many ducks are also kept for show, as pets, or for their ornamental value. Almost all varieties of domesticated duck are descended from the mallard (*Anas platyrhynchos*), apart from the Muscovy duck (*Cairina moschata*).

## Farming

A duck farm in Taiwan

Ducks have been farmed for thousands of years, possibly starting in Southeast Asia. In the Western world, they are not as popular as the chicken, because chickens have much more white lean meat and are easier to keep confined, making the total price much lower for chicken meat, whereas duck is comparatively expensive. While popular in *haute cuisine*, duck appears less frequently in the mass-market food industry and restaurants in the lower price range. However, ducks are more popular in China and there they are raised extensively.

Ducks are farmed for their meat, eggs, and down. A minority of ducks are also kept for *foie gras* production. In Vietnam, their blood is used in a food called *tiết canh*. Their eggs are blue-green to white, depending on the breed.

Ducks can be kept free range, in cages, in barns, or in batteries. Ducks enjoy access to swimming water, but do not require it to survive. They should be fed a grain and insect diet. It is a popular misconception that ducks should be fed bread; bread has limited nutritional value and can be deadly when fed to developing ducklings. Ducks should be monitored for avian influenza, as they are especially prone to infection with the dangerous H5N1 strain.

The females of many breeds of domestic ducks are unreliable at sitting their eggs and raising their young. Notable exceptions include the Rouen duck and especially the Muscovy duck. It has been a custom on farms for centuries to put duck eggs under broody hens for hatching; nowadays incubators are often used. However, young ducklings rely on their mothers for a supply of preen oil to make them waterproof; a chicken hen does not make as much preen oil as a female duck, and an incubator makes none. Once the duckling grows its own feathers, it will produce preen oil from the sebaceous gland near the base of its tail.

## Pets and Ornamentals

Domestic duckling

A domestic Muscovy drake

Domesticated ducks can be kept as pets, in a garden or backyard, and with special accessories, have also been known to be kept in the house. They will often eat insects and slugs. A pond or deep water dish is recommended. If they are given access to a pond, they will dabble in the mud, dredging out and eating wildlife and frog spawn, and swallow adult frogs and toads up to the size of the British common frog *Rana temporaria*, as they have been bred to be much bigger than wild ducks, with a "hull length" (base of neck to base of tail) of up to 1 foot (30 cm) or more; the wild mallard's "hull length" is about 6 inches (15 cm). A coop should be provided for shelter from predators such as foxes, hawks, coyotes, and raccoons, as many breeds of domestic ducks cannot fly.

Ducks are also kept for their ornamental value. Breeds have been developed with crests and tufts or striking plumage. Exhibition shows are held in which ducks, along with other breeds of poultry, are exhibited in competition. These shows can be "open" (meaning any exhibitor who pays the required entry fee can enter), or “closed” (accessible only to members of a given group).

## Aylesbury Duck

Prize-winning Aylesbury duck (front) and drake (rear), 1873

The Aylesbury duck is a breed of domesticated duck, bred mainly for its meat and appearance. It is a large duck with pure white plumage, a pink bill, orange legs and feet, an unusually large keel, and a horizontal stance with its body parallel to the ground. The precise origins of the breed are unclear, but raising white ducks became popular in Aylesbury, Buckinghamshire, England, in the 18th century owing to the demand for white feathers as a filler for quilts. Over the 19th century selective breeding for size, shape and colour led to the Aylesbury duck.

Duck rearing became a major industry in Aylesbury in the 19th century. The ducks were bred on farms in the surrounding countryside. Fertilised eggs were brought into the town’s “Duck End”, where local residents would rear the ducklings in their homes. The opening of a railway to Aylesbury in 1839 enabled cheap and quick transport to the markets of London, and duck rearing became highly profitable. By the 1860s the duck rearing industry began to move out of Aylesbury into the surrounding towns and villages, and the industry in Aylesbury itself began to decline.

In 1873 the Pekin duck was introduced to the United Kingdom. Although its meat was thought to have a poorer flavour than that of the Aylesbury duck, the Pekin was hardier and cheaper to raise. Many breeders switched to the Pekin duck or to Aylesbury-Pekin crosses. By the beginning of the 20th century competition from the Pekin duck, inbreeding, and disease in the pure-bred Aylesbury strain and the rising cost of duck food meant the Aylesbury duck industry was in decline.

The First World War badly damaged the remaining duck industry in Buckinghamshire, wiping out the small scale producers and leaving only a few large farms. Disruption caused by the Second World War further damaged the industry. By the 1950s only one significant flock of Aylesbury ducks remained in Buckinghamshire, and by 1966 there were no duck-breeding or -rearing businesses of any size remaining in Aylesbury itself. Although there is only one surviving flock of pure Aylesbury ducks in the United Kingdom and the breed is critically endangered in the United States, the Aylesbury duck remains a symbol of the town of Aylesbury, and appears on the coat of arms of Aylesbury and on the club badge of Aylesbury United.

## Origins and Description

The precise origin of the Aylesbury duck is unclear. Before the 18th century, duck breeds were rarely recorded in England, and the common duck, bred for farming, was a domesticated form of the wild mallard. The common duck varied in colour, and as in the wild, white ducks would occasionally occur. White ducks were particularly prized, as their feathers were popular as a filler for quilts.

In the 18th century selective breeding of white common ducks led to a white domestic duck, generally known as the English White. Since at least the 1690s ducks had been farmed in Aylesbury, and raising English Whites became popular in Aylesbury and the surrounding villages. By 1813 it was remarked that "ducks form a material article at market from Aylesbury and places adjacent: they are white, and as it seems of an early breed: they are bred and brought up by poor people, and sent to London by the weekly carriers". The duck farmers of Aylesbury went to great lengths to ensure the ducks retained their white colouring, keeping them clear of dirty water, soil with a high iron content and bright sunlight, all of which could discolour the ducks' feathers. Over time, selective breeding of the English White for size and colour gradually led to the development of the Aylesbury duck.

A rather large duck breed, the Aylesbury duck has pure white plumage and bright orange legs and feet. Its legs are placed midway along the body and it stands with its underside parallel to the ground, giving it a body described as "boat-shaped". It has a relatively long and thin swan-like neck, and a long pink bill which comes straight out from the head.[note 1]

An Aylesbury duckling incubates in the egg for 28 days. Until eight weeks after hatching, the time of their first moult, ducks and drakes (females and males) are almost indistinguishable. After moulting, males have two or three curved tail feathers and a fainter, huskier quack than the female. By one year of age, females and males grow to an average weight of 6 and 7 pounds (2.7 and 3.2 kg) respectively, although males can reach around 10 pounds (4.5 kg).

Unlike the Rouen duck, the other popular meat variety in England in the 19th century, Aylesbury ducks lay eggs from early November. Aylesbury ducks fatten quickly and by eight weeks after hatching weigh up to 5 pounds (2.3 kg), large enough to eat but still young and extremely tender. Consequently, their meat came onto the market from February onwards, after the close of the game season but before the earliest spring chickens were on sale. Rouen ducks, whose mallard-like coloration made them less valuable, lay eggs from early February and take six months to grow large enough to eat. As a consequence, Aylesbury ducks were sold primarily in the spring and summer, and Rouen ducks in the autumn and winter.[note 2]

## Aylesbury Duck Farming

The white Aylesbury duck is, and deservedly, a universal favourite. Its snowy plumage and comfortable comportment make it a credit to the poultry-yard, while its broad and deep breast, and its ample back, convey the assurance that your satisfaction will not cease at its death. In parts of Buckinghamshire, this member of the duck family is bred on an extensive scale; not on plains and commons, however, as might be naturally imagined, but in the abodes of the cottagers. Round the walls of the living-rooms, and of the bedroom even, are fixed rows of wooden boxes, lined with hay; and it is the business of the wife and children to nurse and comfort the feathered lodgers, to feed the little ducklings, and to take the old ones out for an airing. Sometimes the "stock" ducks are the cottager's own property, but it more frequently happens that they are intrusted to his care by a wholesale breeder, who pays him so much *per* score for all ducklings properly raised. To be perfect, the Aylesbury duck should be plump, pure white, with yellow feet, and a flesh coloured beak.

*—Isabella Beeton, Mrs Beeton's Book of Household Management, 1861*

Unlike most livestock farming in England at this time, the duck breeders and duck rearers of Aylesbury formed two separate groups. Stock ducks—i.e., ducks kept for breeding—were kept on farms in the countryside of the Aylesbury Vale, away from the polluted air and water of the town. This kept the ducks healthy, and meant a higher number of fertile eggs.

Stock ducks would be chosen from ducklings hatched in March, with a typical breeder keeping six males and twenty laying females at any given time. The females would be kept for around a year before mating, typically to an older male. They would then generally be replaced, to reduce the problems of inbreeding. Stock ducks were allowed to roam freely during the day, and would swim in local ponds which, although privately owned, were treated as common property among the duck breeders; breeders would label their ducks with markings on the neck or head. The stock ducks would forage for greenery and insects, supplemented by greaves (the residue left after the rendering of animal fat). As ducks lay their eggs at night, the ducks would be brought indoors overnight.

Stock Aylesbury ducks at Weston Turville, *circa* 1900

Female Aylesbury ducks would not sit still for the 28 days necessary for their eggs to hatch, and as a consequence the breeders would not allow mothers to sit on their own eggs. Instead the fertilised eggs would be collected and transferred to the "duckers" of Aylesbury's Duck End.[note 3]

## Rearing

The duckers of Aylesbury would buy eggs from the breeders, or be paid by a breeder to raise the ducks on their behalf, and would raise the ducklings in their homes between November and August as a secondary source of income. Duckers were typically skilled labourers, who invested surplus income in ducklings. Many of the tasks related to rearing the ducks would be carried out by the women of the household, particularly the care of newly hatched ducklings.

The eggs would be divided into batches of 13, and placed under broody chickens.[note 4] In the last week of the four-week incubation period the eggs would be sprinkled daily with warm water to soften the shells and allow the ducklings to hatch.

Newly hatched Aylesbury ducklings are timid and thrive best in small groups, so the duckers would divide them into groups of three or four ducklings, each accompanied by a hen. As the ducklings grew older and gained confidence, they would be kept in groups of around 30. Originally the ducks would be kept in every room in the ducker's cottage, but towards the end of the 19th century they were kept in outdoor pens and sheds with suitable protection against cold weather.

The aim of the ducker was to get every duckling as fat as possible by the age of eight weeks (the first moult, the age at which they would be killed for meat), while avoiding any foods which would build up their bones or make their flesh greasy. In their first week after hatching, the ducklings would be fed on boiled eggs, toast soaked in water, boiled rice and beef liver. From the second week on, this diet would gradually be replaced by barley meal and boiled rice mixed with greaves. (Some larger-scale duckers would boil a horse or sheep and feed this to the ducklings in place of greaves.) This high-protein diet was supplemented with nettles, cabbage and lettuce to provide a source of vitamins.[note 5] As with all poultry, ducks require grit in their diet to break up the food and make it digestible. Aylesbury ducklings' drinking water was laced with grit from Long Marston and Gubblecote; this grit also gave their bills their distinctive pinkish colour. Around 85% of ducklings would survive this eight-week rearing process to be sent to market.

It was a very pretty sight to see a flock of young ducklings driven along the village streets to have their one and only swim on a pond; which, taken at the right time, helped them to feather properly. Often the flock spread over the road from side to side and all traffic had perforce to stop, while, quacking much and ambulating slowly, they passed and the road would then be free.

*Walter Rose describing ducks in Haddenham, circa 1925*

While ducks are naturally aquatic, swimming can be dangerous to young ducklings, and it can also restrict a duck's growth. Thus, although duckers would ensure the ducklings always had a trough or sink to paddle in, the ducklings would be kept away from bodies of water while they were growing. The exception was shortly before slaughter, when the ducklings would be taken for one swim in a pond, as it helped them to feather properly.

Although there were a few large-scale duck rearing operations in Aylesbury, raising thousands of ducklings each season, the majority of Aylesbury's duckers would raise between 400 and 1,000 ducklings each year. Because ducking was a secondary occupation, it was not listed in Aylesbury's census returns or directories and it is impossible to know how many people were engaged in it at any given time. Kelly's Directory for 1864 does not list a single duck farmer in Aylesbury, but an 1885 book comments that:

In the early years of the present [19th] century almost every householder at the "Duck End" of the town followed the avocation of ducker. In a living room it was no uncommon sight to meet with young ducks of different ages, divided in pens and monopolizing the greatest space of the apartment, whilst expected new arrivals often were carefully lodged in the bedchamber.

One poor man whom I visited, had before his door a small pit of water, about three yards long and one yard broad: at two corners of this pit are places of shelter for the ducks, thatched with straw: at night the ducks are taken into a house. In one room belonging to this man (the only room he had to live in), were ducks of three growths, on the 14th of January, 1808, fattening for London market: at one corner about seventeen or eighteen four weeks old; at another corner a brood a fortnight old; and at a third corner a brood a week old. In the bed-room were hens brooding ducks' eggs in boxes, to be bought off at different periods.

*The Rev. Richard Parkinson St. John Priest, Secretary to the Norfolk Agricultural Society, reporting on Aylesbury to the Board of Agriculture and Internal Improvement, 1813*

The Duck End was one of the poorer districts of Aylesbury. Until the end of the 19th century it had no sewers or refuse collections. The area had a number of open ditches filled with stagnant water, and outbreaks of malaria and cholera were common. The cottages had inadequate ventilation and lighting, and no running water. Faeces from the duck ponds permeated the local soil and seeped into the cottages through cracks in the floors.

## Slaughter and Sale

When the ducklings were ready for slaughter, the duckers would generally kill them on their own premises. The slaughter would generally take place in the morning, to ensure the ducks would be ready for market in the evening. To keep the meat as white as possible, the ducks would be suspended upside down and their necks broken backwards, and held in this position until their blood had run towards their heads. They were kept in this position for ten minutes before being plucked, as otherwise their blood would collect in those parts of the body from which the feathers had been plucked. The plucking was generally carried out by the women of the household. The plucked carcasses would be sent to market, and the feathers would be sold direct to London dealers.

The market for duck meat in Aylesbury itself was small, and the ducks were generally sent to London for sale. By the 1750s Richard Pococke recorded that four cartloads of ducks were sent from Aylesbury to London every Saturday, and in the late 18th and early 19th centuries the ducks continued to be sent over the Chiltern Hills to London by packhorse or cart.

On 15 June 1839 the entrepreneur and former Member of Parliament for Buckingham, Sir Harry Verney, 2nd Baronet, opened the Aylesbury Railway. Built under the direction of Robert Stephenson, it connected the London and Birmingham Railway's Cheddington railway station on the West Coast Main Line to Aylesbury High Street railway station in eastern Aylesbury. On 1 October 1863 the Wycombe Railway also built a line to Aylesbury, from Princes Risborough railway station to a station on the western side of Aylesbury (the present-day Aylesbury railway station). The arrival of the railway had a powerful impact on the duck industry, and up to a ton of ducks in a night were being shipped from Aylesbury to Smithfield Market in London by 1850.

A routine became established in which salesmen would provide the duckers with labels. The duckers would mark their ducklings with the labels of the firm to whom they wished them to be sold in London. The railway companies would collect ducklings, take them to the stations, ship them to London and deliver them to the designated firms, in return for a flat fee per bird. By avoiding the need for the duckers to travel to market, or the London salesmen to collect the ducklings, this arrangement benefited all concerned, and ducking became very profitable. By 1870 the duck industry was bringing over £20,000 per year into Aylesbury; a typical ducker would make a profit of around £80–£200 per year.[note 6]

## Developments in the Late 19th Century

In 1845, the first National Poultry Show was held, at the Zoological Gardens in London; one of the classes of poultry exhibited was "Aylesbury or other white variety". The personal interest of Queen Victoria in poultry farming, and its inclusion in the Great Exhibition of 1851, further raised public interest in poultry. From 1853 the Royal Agricultural Society and the Bath and West of England Society, the two most prominent agricultural societies in England, included poultry sections in their annual agricultural shows. This in turn caused smaller local poultry shows to develop across the country.

Breeders would choose potential exhibition ducks from among newly hatched ducklings in March and April, and they would be given a great deal of extra attention. They would be fed a carefully controlled diet to get them to the maximum weight, and would be allowed out for a few hours each day to keep them in as good a physical condition as possible. Before the show, their legs and feet would be washed, their bills trimmed with a knife and sandpapered smooth, and their feathers brushed with linseed oil. While most breeders would give the ducks a healthy meal before the show to calm them, some breeders would force-feed the ducks with sausage or worms, to get them to as heavy a weight as possible. Exhibition standards judged an Aylesbury duck primarily on size, shape and colour. This encouraged the breeding of larger ducks, with pronounced exaggerated keels, and loose baggy skin. By the beginning of the 20th century the Aylesbury duck had diverged into two separate strains, one bred for appearance and one for meat.

## Pekin Ducks

Pekin ducks are similar to the Aylesbury, but with orange bills and a more upright posture.

In 1873 the Pekin duck was introduced from China to Britain for the first time. Superficially similar in appearance to an Aylesbury duck, a Pekin is white with orange legs and bill, with its legs near the rear, giving it an upright stance while on land. Although not thought to have such a delicate flavour as the Aylesbury, the Pekin was hardier, a more prolific layer, fattened more quickly, and was roughly the same size as an Aylesbury at nine weeks.

Aylesbury ducks, meanwhile, were becoming inbred, meaning fertile eggs were scarcer and the ducks were more susceptible to disease. Exhibition standards had led to selection for an exaggerated keel by breeders, despite it being unpopular with dealers and consumers. Poultry show judges also admired the long neck and upright posture of Pekin ducks over the boat-like stance of the Aylesbury. Some of the breeders in the Aylesbury area began to cross Pekin ducks with the pure Aylesbury strain. Although the Aylesbury-Pekin cross ducks did not have the delicate flavour of the pure Aylesbury, they were hardier and much cheaper to raise.

Until the mid-19th century duck rearing was concentrated on the Duck End, but by the 1860s it had spread to many other towns and villages in the area, particularly Weston Turville and Haddenham. Contamination of Aylesbury's soil by years of duck rearing, and new public health legislation which ended many traditional practices, caused the decline of the duck rearing industry in the Duck End, and by the 1890s the majority of Aylesbury ducks were raised in the villages rather than the town itself. Population shifts and the improved national rail network reduced the need to rear ducks near London, and large duck farms opened in Lancashire, Norfolk and Lincolnshire. Although the number of ducks raised nationwide continued to grow, between 1890 and 1900 the number of ducks raised in the Aylesbury area remained static, and from 1900 it began to drop.

## Decline

Jemima Puddle-Duck

By the time Beatrix Potter's 1908 *The Tale of Jemima Puddle-Duck*—about an Aylesbury duck although set in Cumbria—caused renewed interest in the breed, the Aylesbury duck was in steep decline. The duckers of Buckinghamshire had generally failed to introduce technological improvements such as the incubator, and inbreeding had dangerously weakened the breed. Meanwhile, the cost of duck food had risen fourfold over the 19th century, and from 1873 onwards competition from Pekin and Pekin cross ducks was undercutting Aylesbury ducks at the marketplace.

The First World War devastated the remaining duckers of Buckinghamshire. The price of duck food rose steeply while the demand for luxury foodstuffs fell, and wartime restructuring ended the beneficial financial arrangements with the railway companies. By the end of the war small-scale duck rearing in the Aylesbury Vale had vanished, with duck raising dominated by a few large duck farms. Shortages of duck food in the Second World War caused further disruption to the industry, and almost all duck farming in the Aylesbury Vale ended. A 1950 "Aylesbury Duckling Day" campaign to boost the reputation of the Aylesbury duck had little effect; by the end of the 1950s the last significant farms had closed, other than a single flock in Chesham owned by Mr L. T. Waller, and by 1966 there were no duck breeders or rearers of any size remaining in Aylesbury. As of 2015 the Waller family's farm in Chesham remains in business, the last surviving flock of pure Aylesbury meat ducks in the country.

Aylesbury ducks were imported into the United States in 1840, although they never became a popular breed. They were, however, added to the American Poultry Association's *Standard of Perfection* breeding guidelines in 1876. As of 2013, the breed was listed as critically endangered in the United States by The Livestock Conservancy.

## Legacy

The Aylesbury duck remains a symbol of the town of Aylesbury. Aylesbury United F.C. are nicknamed "The Ducks" and include an Aylesbury duck on their club badge, and the town's coat of arms includes an Aylesbury duck and plaited straw, representing the two historic industries of the town. The Aylesbury Brewery Company, now defunct, featured the Aylesbury duck as its logo, an example of which can still be seen at the Britannia pub. Duck Farm Court is a shopping area of modern Aylesbury located near the historic hamlet of California, close to one of the main breeding grounds for ducks in the town, and there have been two pubs in the town with the name "The Duck" in recent years; one in Bedgrove that has since been demolished and one in Jackson Road that has recently been renamed.

# Indian Runner Duck

Indian Runners are an unusual breed of *Anas platyrhynchos domesticus*, the domestic duck. They stand erect like penguins and, rather than waddling, they run. The females usually lay about 150 – 200 eggs a year or more, depending whether they are from exhibition or utility strains. They were found on the Indonesian islands of Lombok, Java and Bali where they were 'walked' to market and sold as egg-layers or for meat. These ducks do not fly and only rarely form nests and incubate their own eggs. They run or walk, often dropping their eggs wherever they happen to be. Duck-breeders need to house their birds overnight or be vigilant in picking up the eggs to prevent them from being taken by other animals.

The ducks vary in weight between 1.4 and 2.3 kg (3-5 lbs). Their height (from crown to tail tip) ranges from 50 cm (20 inches) in small females to about 76 cm (30 inches) in the taller males. The eggs are often greenish-white in color, but these too vary. Drakes have a small curl on the tip of their tails, while hens have flat tails. It's difficult to determine their sex until they are fully mature.

They often swim in ponds and streams, but they are likely to be preoccupied foraging in grassy meadows for worms, slugs, even catching flies. They appreciate open spaces but are happy in gardens from which they cannot fly and where they make much less noise than call ducks. Only females quack and drakes are limited to a hoarse whisper. Compared to big table ducks, they eat less grain and pellet supplements. When they are young, they eat rocks to form their gizzard which helps them digest their food. They do not have teeth, so they must have water to soften food.

Indian Runner ducks

## Origins of the Breed

The Indian Runner ducks are domesticated waterfowl that live in the archipelago of the East Indies. There is no evidence that they came originally from India itself. Attempts by British breeders at the beginning of the twentieth century to find examples in the subcontinent had very limited success. Like many other breeds of waterfowl imported into Europe and America, the term 'Indian' may well be fanciful, denoting a loading port or the transport by 'India-men' sailing ships of the East India Company. Other misnamed geese and ducks include the African goose, the black East Indian duck and the Muscovy duck.

The Runner became popular in Europe and America as an egg-laying variety towards the end of the nineteenth century largely as a result of an undated pamphlet called *The India Runner: its History and Description* published by John Donald of Wigton between 1885 and 1890. Donald's publication is advertised briefly in *The Feathered World*, 1895, under the title of "The Indian Runner Duck". Donald describes the pied variety and gives the popular story of the importation into Cumbria (Northwest England) by a sea captain some fifty years earlier.

The breed is unusual not only for its high egg production but also for its upright stance and variety of color genes, some of which are seen in seventeenth century Dutch paintings. Other references to such domestic ducks use the names 'Penguin Ducks' and 'Baly Soldiers'. Harrison Weir's *Our Poultry* (1902) describes the Penguin Ducks belonging to Mr Edward Cross in the Surrey Zoological Gardens between 1837-38. These may well have been imported by the 13th Earl of Derby. Darwin describes them (1868) as having elongated 'femur and meta-tarsi', contrary to Tegetmeier's assertions.

The Cumbrian importations, according to Matthew Smith in 1923, included completely fawn Runners and completely white Runners as well as the pied (fawn-and-white and grey-and-white) varieties. The most successful attempt to import fresh blood lines was by Joseph Walton between 1908 and 1909. Accounts of these ventures can be found in Coutts (1927) and Ashton (2002). Walton shipped in birds from Lombok and Java, revolutionizing the breeding stock which, according to Donald, had become badly mixed with local birds. Further importations by Miss Chisholm and Miss Davidson in 1924 and 1926 continued to revive the breed.

## Development

Pure breed enthusiasts, exhibitors and show judges wanted to establish standard descriptions. Standards were drawn up in by the Waterfowl Club in England (1897) and America (1898) for the pied color varieties. These were largely the same until 1915 when the two countries diverged. The American Poultry Association chose a variety with blue in the genotype whilst the English Poultry Club Standard kept to the pure form described by Donald in his original pamphlet. Other colors followed making use of black genes brought in by some of Walton's birds. These were to produce black, chocolate and Cumberland blue. Later were developed the mallard, trout, blue trout, and apricot trout versions. Slightly different names and descriptions can be found in American and German standards. An account of the influence of the Indian Runner Duck Club (founded in 1906), particularly the input by John Donald, Joseph Walton, Dr J.A. Coutts and Matthew Smith, can be found in Ashton (2002).

The most profound impact of the Indian Runners was on the development of the modern 'light duck' breeds. Before 1900, most ducks were bred for the table. Aylesbury and Rouen ducks were famous throughout the nineteenth century, and these were supplemented or replaced, after 1873-4, by importation from China of the Pekin duck. As soon as the Indian Runners became fashionable, a demand for egg-layers and general purpose breeds developed. Using Runners crossed to Rouens, Aylesburys and Cayugas (the large black American breed), William Cook produced his famous Orpington Ducks. Mrs Campbell crossed her fawn-and-white Runner duck to a Rouen drake to create the Campbell ducks introduced in 1898. Later, she introduced wild mallard blood and managed to create the most prolific egg-layer, the Khaki Campbell (announced in 1901). Other breeds followed, some of which emerged as direct mutations of the Khaki Campbell, along with crosses back to Indian Runners, the most famous being the Abacot Ranger (known in Germany as the Streicher) and the Welsh Harlequin.

## Color Breeding

Indian Runner ducks and Pekins brought in unusual plumage colour mutations. These included the dusky and restricted mallard genes, light phase, harlequin phase, blue and brown dilutions, as well as the famous pied varieties named by the geneticist F.M. Lancaster as the 'Runner pattern'. Much of the proliferation of new colour varieties in breeds of domestic duck begins with the importation of these oriental ducks. Original research by R.G. Jaap (1930s) and F.M. Lancaster has allowed breeders to understand the effect of genotypes in managing and creating colour varieties. Simplified information can be found in writings by Dave Holderread, and Mike and Chris Ashton.

## References


- Roehrig, Catharine H.; Dreyfus, Renée; Keller, Cathleen A. (2005). Hatshepsut: From Queen to Pharaoh. New York: Metropolitan Museum of Art. p. 268. ISBN 978-1-58839-173-5.
- Dohner, Janet Vorwald (2001-01-01). The Encyclopedia of Historic and Endangered Livestock and Poultry Breeds. Yale University Press. ISBN 030013813X.
- Berhardt, Clyde E. B. (1986). I Remember: Eighty Years of Black Entertainment, Big Bands. Philadelphia: University of Pennsylvania Press. p. 153. ISBN 978-0-8122-8018-0. OCLC 12805260.
- Grandin, Temple; Johnson, Catherine (2005). Animals in Translation. New York City: Scribner. pp. 69–71. ISBN 0-7432-4769-8.
- Alan Davidson (11 August 2014), The Oxford Companion to Food, Oxford University Press, p. 389, ISBN 978-0-19-967733-7
- Ekarius, Carol (2007). Storey's Illustrated Guide to Poultry Breeds. 210 MAS MoCA Way, North Adams MA 01247: Storey Publishing. ISBN 978-1-58017-667-5.
- Kiple, Kenneth F.; Ornelas, Kriemhild Coneè (2000). The Cambridge World History of Food. Cambridge, UK: Cambridge University Press. ISBN 0-521-40214-X. OCLC 44541840.
- Herrema, Martin (12 December 2014). "Chickens 'closer to dinosaurs' than other birds". University of Kent. Retrieved 20 June 2016.
- Paul Sheridan (November 8, 2015). "The Sacred Chickens of Rome". anecdotesfromantiquity.net. Retrieved November 17, 2015.

# 6

# Poultry Products

Poultry products are used mainly for food. Some of the chapters included in this section are chicken as food, duck as food, feather meal, egg as food and down feather. The chapter closely examines all the poultry products that are used in day-to-day life.

## Egg as Food

Eggs are laid by female animals of many different species, including birds, reptiles, amphibians, mammals, and fish, and have been eaten by humans for thousands of years. Bird and reptile eggs consist of a protective eggshell, albumen (egg white), and vitellus (egg yolk), contained within various thin membranes. The most popular choice for egg consumption are chicken eggs. Other popular choices for egg consumption are duck, quail, roe, and caviar.

A brown and a white chicken egg1

Egg yolks and whole eggs store significant amounts of protein and choline, and are widely used in cookery. Due to their protein content, the United States Department of Agriculture categorizes eggs as *Meats* within the Food Guide Pyramid. Despite the nutritional value of eggs, there are some potential health issues arising from egg quality, storage, and individual allergies.

Chickens and other egg-laying creatures are widely kept throughout the world, and mass production of chicken eggs is a global industry. In 2009, an estimated 62.1 million metric tons of eggs were produced worldwide from a total laying flock of approximately 6.4 billion hens. There are issues of regional variation in demand and expectation, as well as current debates concerning methods of mass production. The European Union recently banned battery husbandry of chickens.

A cracked raw chicken egg

## History

Bird eggs have been valuable foodstuffs since prehistory, in both hunting societies and more recent cultures where birds were domesticated. The chicken was probably domesticated for its eggs from jungle fowl native to tropical and subtropical Southeast Asia and India before 7500 BCE. Chickens were brought to Sumer and Egypt by 1500 BCE, and arrived in Greece around 800 BCE, where the quail had been the primary source of eggs. In Thebes, Egypt, the tomb of Haremhab, built about 1420 BCE, shows a depiction of a man carrying bowls of ostrich eggs and other large eggs, presumably those of the pelican, as offerings. In ancient Rome, eggs were preserved using a number of methods, and meals often started with an egg course. The Romans crushed the shells in their plates to prevent evil spirits from hiding there. In the Middle Ages, eggs were forbidden during Lent because of their richness. The word mayonnaise possibly was derived from *moyeu*, the medieval French word for the yolk, meaning center or hub.

Egg scrambled with acidic fruit juices were popular in France in the 17th century; this may have been the origin of lemon curd.

The dried egg industry developed in the 19th century, before the rise of the frozen egg industry. In 1878, a company in St. Louis, Missouri started to transform egg yolk and white into a light-brown, meal-like substance by using a drying process. The production of dried eggs significantly expanded during World War II, for use by the United States Armed Forces and its allies.

In 1911, the egg carton was invented by Joseph Coyle in Smithers, British Columbia, to solve a dispute about broken eggs between a farmer in Bulkley Valley and the owner of the Aldermere Hotel. Early egg cartons were made of paper.

## Varieties

Bird eggs are a common food and one of the most versatile ingredients used in cooking. They are important in many branches of the modern food industry.

Quail eggs (upper left), chicken egg (lower left) and ostrich egg (right)

The most commonly used bird eggs are those from the chicken. Duck and goose eggs, and smaller eggs, such as quail eggs, are occasionally used as a gourmet ingredient in western countries. Eggs are a common everyday food in many parts of Asia, such as China and Thailand, with Asian production providing 59% of the world total in 2013.

The largest bird eggs, from ostriches tend to be used only as special luxury food. Gull eggs are considered a delicacy in England, as well as in some Scandinavian countries, particularly in Norway. In some African countries, guineafowl eggs are commonly seen in marketplaces, especially in the spring of each year. Pheasant eggs and emu eggs are edible, but less widely available. Sometimes they are obtainable from farmers, poulterers, or luxury grocery stores. In many countries, wild birds' eggs are protected by laws which prohibit collecting or selling them, or permit collection only during specific periods of the year.

| Global production of chicken eggs (in shell) (2013, in millions of tonnes) | |
|---|---|
| China | 24.8 |
| USA | 5.6 |
| India | 3.8 |
| Japan | 2.5 |
| **World** | **68.3** |
| Source: United Nations, Food and Agriculture Organization, FAOSTAT | |

## Production

In 2013, world production of chicken eggs was 68.3 million tonnes, with China contributing 37% of this total.

## Anatomy and Characteristics

A raw chicken egg within its membrane, the shell removed by soaking in vinegar

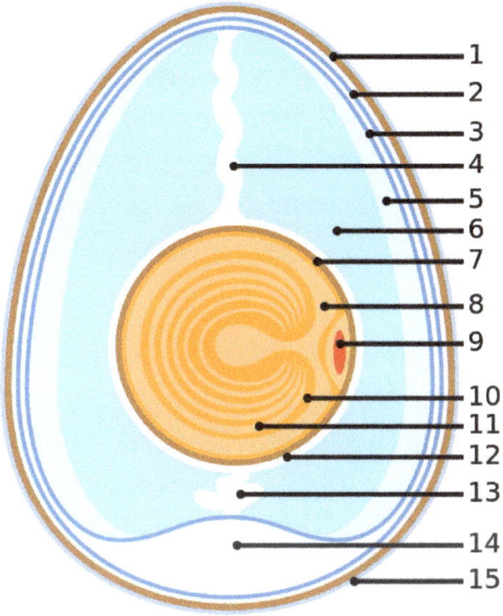


Schematic of a chicken egg:

1. Eggshell
2. Outer membrane
3. Inner membrane
4. Chalaza
5. Exterior albumen
6. Middle albumen
7. Vitelline membrane
8. Nucleus of pander
9. Germinal disc (nucleus)
10. Yellow yolk
11. White yolk
12. Internal albumen
13. Chalaza
14. Air cell
15. Cuticula

The shape of an egg resembles a prolate spheroid with one end larger than the other, with cylindrical symmetry along the long axis.

An egg is surrounded by a thin, hard shell. Inside, the egg yolk is suspended in the egg white by one or two spiral bands of tissue called the chalazae.

## Air Cell

The larger end of the egg contains the air cell that forms when the contents of the egg cool down and contract after it is laid. Chicken eggs are graded according to the size of this air cell, measured during candling. A very fresh egg has a small air cell and receives a grade of AA. As the size of the air cell increases and the quality of the egg decreases, the grade moves from AA to A to B. This provides a way of testing the age of an egg: as the air cell increases in size due to air being drawn through pores in the shell as water is lost, the egg becomes less dense and the larger end of the egg will rise to increasingly shallower depths when the egg is placed in a bowl of water. A very old egg will actually float in the water and should not be eaten.

## Shell

Egg shell color is caused by pigment deposition during egg formation in the oviduct and can vary according to species and breed, from the more common white or brown to pink or speckled blue-green. In general, chicken breeds with white ear lobes lay white eggs, whereas chickens with red ear lobes lay brown eggs. Although there is no significant link between shell color and nutritional value, there is often a cultural preference for one color over another.

## Membrane

The membrane is a clear film lining the eggshell, visible when one peels a boiled egg. Eggshell membrane is primarily composed of fibrous proteins such as collagen type I.

## White

White is the common name for the clear liquid (also called the albumen or the glair/glaire) contained within an egg. In chickens it is formed from the layers of secretions of the anterior section of the hen's oviduct during the passage of the egg. It forms around either fertilized or unfertilized yolks. The primary natural purpose of egg white is to protect the yolk and provide additional nutrition for the growth of the embryo.

Egg white consists primarily of about 90% water into which is dissolved 10% proteins (including albumins, mucoproteins, and globulins). Unlike the yolk, which is high in lipids (fats), egg white contains almost no fat and the carbohydrate content is less than

1%. Egg white has many uses in food, and many others, including the preparation of vaccines such as those for influenza.

## Yolk

The yolk in a newly laid egg is round and firm. As the yolk ages, it absorbs water from the albumen, which increases its size and causes it to stretch and weaken the vitelline membrane (the clear casing enclosing the yolk). The resulting effect is a flattened and enlarged yolk shape.

Yolk color is dependent on the diet of the hen; if the diet contains yellow/orange plant pigments known as xanthophylls, then they are deposited in the yolk, coloring it. Lutein is the most abundant pigment in egg yolk. A colorless diet can produce an almost colorless yolk. Yolk color is, for example, enhanced if the diet includes products such as yellow corn and marigold petals. In the US, the use of artificial color additives is forbidden.

## Abnormalities

Shell-less or thin-shelled eggs can be caused by egg drop syndrome.

## Culinary Properties

## Types of Dishes

A fried chicken egg, "sunny side up"

Chicken eggs are widely used in many types of dishes, both sweet and savory, including many baked goods. Some of the most common preparation methods include scrambled, fried, hard-boiled, soft-boiled, omelettes and pickled. They can also be eaten raw, though this is not recommended for people who may be especially susceptible to salmonellosis, such as the elderly, the infirm, or pregnant women. In addition, the protein in raw eggs is only 51% bioavailable, whereas that of a cooked egg is nearer 91% bio-available, meaning the protein of cooked eggs is nearly twice as absorbable as the protein from raw eggs.

As an ingredient, egg yolks are an important emulsifier in the kitchen, and are also used as a thickener in custards.

Soft-boiled quail eggs, with potato galettes

The albumen, or egg white, contains protein, but little or no fat, and can be used in cooking separately from the yolk. The proteins in egg white allow it to form foams and aerated dishes. Egg whites may be aerated or whipped to a light, fluffy consistency, and are often used in desserts such as meringues and mousse.

Ground egg shells are sometimes used as a food additive to deliver calcium. Every part of an egg is edible, although the eggshell is generally discarded. Some recipes call for immature or unlaid eggs, which are harvested after the hen is slaughtered or cooked while still inside the chicken.

## Cooking

A few tips for baking with eggs and cakes.

Eggs contain multiple proteins which gel at different temperatures within the yolk and the white, and the temperature determines the gelling time. Egg yolk begins to gelify, or solidify, when it reaches temperatures between about 60 and 70 °C (140 and 158 °F).

Egg white gels at slightly higher temperatures, about 60 to 80 °C (140 to 176 °F)- the white contains ovalbumin that sets at the highest temperature. However, in practice, in many cooking processes the white gels first because it is exposed to higher temperatures for longer.

half boiled egg dish

Salmonella is killed instantly at 71 °C (160 °F), but is also killed from 54.5 °C (130.1 °F) if held there for sufficiently long time periods. To avoid the issue of salmonella, eggs can be pasteurized in-shell at 57 °C (135 °F) for an hour and 15 minutes. Although the white is slightly milkier, the eggs can be used in normal ways. Whipping for meringue takes significantly longer, but the final volume is virtually the same.

If a boiled egg is overcooked, a greenish ring sometimes appears around egg yolk due to the iron and sulfur compounds in the egg. It can also occur with an abundance of iron in the cooking water. The green ring does not affect the egg's taste; overcooking, however, harms the quality of the protein. Chilling the egg for a few minutes in cold water until it is completely cooled may prevent the greenish ring from forming on the surface of the yolk.

Peeling a cooked egg is easiest when the egg was put into boiling water as opposed to slowly heating the egg in cold water.

## Flavor Variations

A batch of tea eggs with shell intact soaking in a brew of spices and tea

Although the age of the egg and the conditions of its storage have a greater influence, the bird's diet does affect the flavor of the egg. For example, when a brown-egg chicken breed eats rapeseed or soy meals, its intestinal microbes metabolize them into fishy-smelling triethylamine, which ends up in the egg. The unpredictable diet of free-range hens will produce unpredictable eggs. Duck eggs tend to have a flavor distinct from, but still resembling chicken eggs.

Eggs can also be soaked in mixtures to absorb flavor. Tea eggs are steeped in a brew from a mixture of various spices, soy sauce, and black tea leaves to give flavor.

## Storage

Careful storage of edible eggs is extremely important, as an improperly handled egg can contain elevated levels of *Salmonella* bacteria that can cause severe food poisoning. In the US, eggs are washed, and this cleans the shell, but erodes the cuticle. The USDA thus recommends refrigerating eggs to prevent the growth of *Salmonella*.

Refrigeration also preserves the taste and texture. However, intact eggs can be left unrefrigerated for several months without spoiling. In Europe, eggs are not usually washed, and the shells are dirtier, however the cuticle is undamaged, and they do not require refrigeration. In the UK in particular, hens are immunized against salmonella, and the eggs are generally safe for 21 days.

## Preservation

Salted duck egg

The simplest method to preserve an egg is to treat it with salt. Salt draws water out of bacteria and molds, which prevents their growth. The Chinese salted duck egg is made by immersing duck eggs in brine, or coating them individually with a paste of salt and mud or clay. The eggs stop absorbing salt after about a month, having reached osmotic equilibrium. Their yolks take on an orange-red color and solidify, but the white remains somewhat liquid. They are often boiled before consumption, and are often served with rice congee.

Pickled egg, colored with beetroot juice

Another method is to make pickled eggs, by boiling them first and immersing them in a mixture of vinegar, salt and spices, such as ginger or allspice. Frequently, beetroot juice is added to impart a red color to the eggs. If the eggs are immersed in it for a few hours, the distinct red, white, and yellow colors can be seen when the eggs are sliced. If marinated for several days or more, the red color will reach the yolk. If the eggs are marinated in the mixture for several weeks or more, the vinegar will dissolve much of the shell's calcium carbonate and penetrate the egg, making it acidic enough to inhibit the growth of bacteria and molds. Pickled eggs made this way will generally keep for a year or more without refrigeration.

Century egg

A century egg or hundred-year-old egg is preserved by coating an egg in a mixture of clay, wood ash, salt, lime, and rice straw for several weeks to several months, depending on the method of processing. After the process is completed, the yolk becomes a dark green, cream-like substance with a strong odor of sulfur and ammonia, while the white becomes a dark brown, transparent jelly with a comparatively mild, distinct flavor. The transforming agent in a century egg is its alkaline material, which gradually raises the pH of the egg from around 9 to 12 or more. This chemical process breaks down some of the complex, flavorless proteins and fats of the yolk into simpler, flavorful ones, which in some way may be thought of as an "inorganic" version of fermentation.

## Cooking Substitutes

For those who do not consume eggs, alternatives used in baking include other rising agents or binding materials, such as ground flax seeds or potato starch flour. Tofu can also act as a partial binding agent, since it is high in lecithin due to its soy content. Applesauce can be used, as well as arrowroot and banana. Extracted soybean lecithin, in turn, is often used in packaged foods as an inexpensive substitute for egg-derived lecithin. Chickpea brine, also known as aquafaba, can replace egg whites in desserts like meringues and mousses.

Other egg substitutes are made from just the white of the egg for those who worry about the high cholesterol and fat content in eggs. These products usually have added vitamins and minerals, as well as vegetable-based emulsifiers and thickeners such as xanthan gum or guar gum. These allow the product to maintain the nutrition and several culinary properties of real eggs, making possible foods such as Hollandaise sauce, custard, mayonnaise, and most baked goods with these substitutes.

## Nutritional Value

Chicken eggs, the most commonly eaten eggs, provide 155 calories (kcal) of food energy and 12.6 g of protein in a 100 gram serving.

Eggs (boiled) supply several vitamins and minerals as significant amounts of the Daily Value (DV), including vitamin A (19% DV), riboflavin (42% DV), pantothenic acid (28% DV), vitamin $B_{12}$ (46% DV), choline (60% DV), phosphorus (25% DV), zinc (11% DV) and vitamin D (15% DV) (table per 100 gram serving of a hard-boiled egg).

A yolk contains more than two-thirds of the recommended daily intake of 300 mg of cholesterol (table).

The diet of laying hens can affect the nutritional quality of eggs. For instance, chicken eggs that are especially high in omega-3 fatty acids are produced by feeding hens a diet containing polyunsaturated fats from sources like fish oil, chia seeds or flaxseeds. Pasture-raised free-range hens, which forage for their own food, also produce eggs that are relatively enriched in omega-3 fatty acids compared to cage-raised chickens.

A 2010 USDA study determined there were no significant differences of macronutrients in various chicken eggs.

Cooked eggs are easier to digest, as well as having a lower risk of salmonellosis.

## Health Effects

### Cholesterol and Fat

More than half the calories found in eggs come from the fat in the yolk; 50 grams of chicken egg (the contents of an egg just large enough to be classified as "large" in the US, but "medium" in Europe) contains approximately 5 grams of fat. People on a low-cholesterol diet may need to reduce egg consumption; however, only 27% of the fat in egg is saturated fat (palmitic, stearic and myristic acids). The egg white consists primarily of water (87%) and protein (13%) and contains no cholesterol and little, if any, fat.

There is debate over whether egg yolk presents a health risk. Some research suggests dietary cholesterol increases the ratio of total to HDL cholesterol and, therefore, adversely affects the body's cholesterol profile; whereas other studies show that moderate consumption of eggs, up to one a day, does not appear to increase heart disease risk in healthy individuals. Harold McGee argues that the cholesterol in the yolk is not what causes a problem, because fat (in particular, saturated) is much more likely to raise cholesterol levels than the actual consumption of cholesterol.

## Type 2 Diabetes

Studies have shown conflicting results about a possible connection between egg consumption and type two diabetes. A 1999 prospective study of over 117,000 people by the Harvard School of Public Health concluded, in part, that "The apparent increased risk of CHD associated with higher egg consumption among diabetic participants warrants further research." A 2008 study by the Physicians' Health Study I (1982–2007) and the Women's Health Study (1992–2007) determined the "data suggest that high levels of egg consumption (daily) are associated with an increased risk of type 2 diabetes." However, a study published in 2010 found no link between egg consumption and type 2 diabetes. A meta-analysis from 2013 found that eating 4 eggs per week increased the risk of diabetes by 29%. Another meta-analysis from 2013 also supported the idea that egg consumption may lead to an increased incidence of type two diabetes.

## Cardiovascular Risk

Eggs are one of the largest sources of phosphatidylcholine (lecithin) in the human diet. A study published in the scientific journal Nature showed that dietary phosphatidylcholine is digested by bacteria in the gut and eventually converted into the compound TMAO, a compound linked with increased heart disease.

The 1999 Harvard School of Public Health study of 37,851 men and 80,082 women concluded that its "findings suggest that consumption of up to 1 egg per day is unlikely to have substantial overall impact on the risk of CHD or stroke among healthy men and women." In a study of 4,000 people, scientists found that eating eggs increased blood levels of a metabolite promoting atherosclerosis, TMAO, and that this in turn caused significantly higher risk of heart attack and stroke after three years of follow-up.

A 2007 study of nearly 10,000 adults demonstrated no correlation between moderate (six per week) egg consumption and cardiovascular disease or strokes, except in the subpopulation of diabetic patients who presented an increased risk of coronary artery disease. One potential alternative explanation for the null finding is that background dietary cholesterol may be so high in the usual Western diet that adding somewhat more has little further effect on blood cholesterol. Other research supports the idea that a high egg intake increases cardiovascular risk in diabetic patients. A 2009 prospective cohort study of over 21,000 individuals suggests that "egg consumption up to 6/week has no major effect on the risk of cardiovascular disease and mortality and that consumption of 7+/week is associated with a modest increased risk of total mortality" in males, whereas among males with diabetes, "any egg consumption is associated with an increased risk of all-cause mortality and there was suggestive evidence for an increased risk of myocardial infarction and stroke". A 2013 meta-analysis found no association between egg consumption and heart disease or stroke. A 2013 systematic review and meta-analysis found no association between egg consumption and cardiovascular disease or cardiovascular disease mortality, but did find egg consumption more than once daily increased cardiovascular disease risk 1.69-fold in those with type 2 diabetes mellitus compared to type 2 diabetics who ate less than 1 egg per week. Another 2013 meta-analysis found that eating 4 eggs per week increased the risk of cardiovascular disease by 6%.

## Contamination

Egg cleaning on a farm in Norway

A health issue associated with eggs is contamination by pathogenic bacteria, such as *Salmonella enteritidis*. Contamination of eggs exiting a female bird via the cloaca may also occur with other members of the *Salmonella* genus, so care must be taken to prevent the egg shell from becoming contaminated with fecal matter. In commercial practice in the US, eggs are quickly washed with a sanitizing solution within minutes of being laid. The risk of infection from raw or undercooked eggs is dependent in part upon the sanitary conditions under which the hens are kept.

Health experts advise people to refrigerate washed eggs, use them within two weeks, cook them thoroughly, and never consume raw eggs. As with meat, containers and surfaces that have been used to process raw eggs should not come in contact with ready-to-eat food.

A study by the U.S. Department of Agriculture in 2002 (Risk Analysis April 2002 22(2):203-18) suggests the problem is not as prevalent as once thought. It showed that of the 69 billion eggs produced annually, only 2.3 million are contaminated with *Salmonella*—equivalent to just one in every 30,000 eggs—thus showing *Salmonella* infection is quite rarely induced by eggs. However, this has not been the case in other countries, where *Salmonella enteritidis* and *Salmonella typhimurium* infections due to egg consumptions are major concerns. Egg shells act as hermetic seals that guard against bacteria entering, but this seal can be broken through improper handling or if laid by unhealthy chickens. Most forms of contamination enter through such weaknesses in the shell. In the UK, the British Egg Industry Council award the lions stamp to eggs that, among other things, come from hens that have been vaccinated against *Salmonella*.

## Food Allergy

One of the most common food allergies in infants is eggs. Infants usually have the opportunity to grow out of this allergy during childhood, if exposure is minimized. Allergic reactions against egg white are more common than reactions against egg yolks.

In addition to true allergic reactions, some people experience a food intolerance to egg whites.

Food labeling practices in most developed countries now include eggs, egg products and the processing of foods on equipment that also process foods containing eggs in a special allergen alert section of the ingredients on the labels.

## Farming Issues

Eggs for sale at a grocery store

White and brown eggs in an egg crate.

Most commercially farmed chicken eggs intended for human consumption are unfertilized, since the laying hens are kept without roosters. Fertile eggs can be eaten, with a little nutritional difference to the unfertilized. Fertile eggs will not contain a developed embryo, as refrigeration temperatures inhibit cellular growth for an extended time. Sometimes an embryo is allowed to develop but eaten before hatching as with balut.

## Grading By Quality and Size

The US Department of Agriculture grades eggs by the interior quality of the egg and the appearance and condition of the egg shell. Eggs of any quality grade may differ in weight (size).

- U.S. Grade AA
    - Eggs have whites that are thick and firm; yolks that are high, round, and practically free from defects; and clean, unbroken shells.
    - Grade AA and Grade A eggs are best for frying and poaching, where appearance is important.
- U.S. Grade A
    - Eggs have characteristics of Grade AA eggs except the whites are "reasonably" firm.
    - This is the quality most often sold in stores.
- U.S. Grade B
    - Eggs have whites that may be thinner and yolks that may be wider and flatter than eggs of higher grades. The shells must be unbroken but may show slight stains.
    - This quality is seldom found in retail stores because they are usually used to make liquid, frozen, and dried egg products, as well as other egg-containing products.

In Australia and the European Union, eggs are graded by the hen farming method, free range, battery caged, etc.

Chicken eggs are also graded by size for the purpose of sales. Some maxi eggs can have double-yolk and some farms separate out double-yolk eggs for special sale.

Comparison of an egg and a maxi egg with a double-yolk - Closed (1/2)

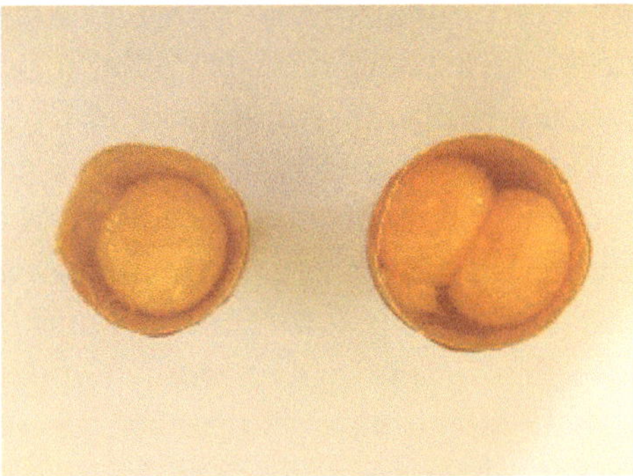

Comparison of an egg and a maxi egg with a double-yolk - Opened (2/2)

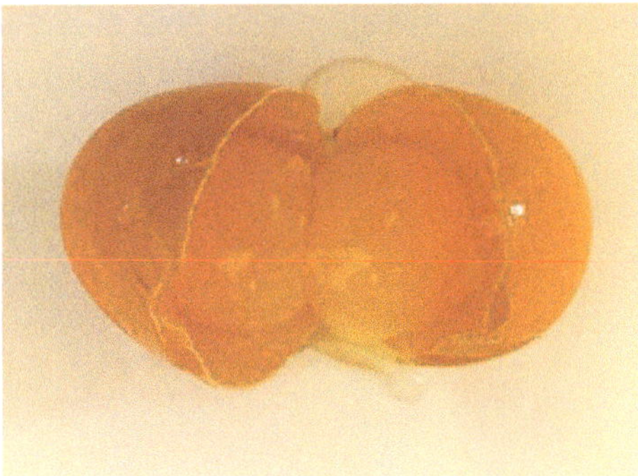

Double-yolk egg - Opened

## Washing and Refrigeration

In North America, the legislation requires eggs to be washed and refrigerated before being sold to consumers. This is to remove natural farm contaminants present in the cleanest farms and to prevent the growth of bacteria. In Europe legislation requires the opposite. Washing removes the natural protective cuticle on the egg and refrigeration causes condensation which may promote bacteria growth.

## Color of Eggshell

White, speckled (red), and brown chicken eggs

Although egg color is a largely cosmetic issue, with no effect on egg quality or taste, it is a major issue in production due to regional and national preferences for specific colors, and the results of such preferences on demand. For example, in most regions of the United States, chicken eggs are generally white. In some parts of the northeast of that country, particularly New England, where a television jingle for years proclaimed "brown eggs are local eggs, and local eggs are fresh!", brown eggs are more common. Local chicken breeds, including the Rhode Island Red, lay brown eggs. Brown eggs are also preferred in Costa Rica, Ireland, France, and the United Kingdom. In Brazil and Poland, white chicken eggs are generally regarded as industrial, and brown or reddish ones are preferred. Small farms and smallholdings, particularly in economically advanced nations, may sell eggs of widely varying colors and sizes, with combinations of white, brown, speckled (red), green, and blue (as laid by certain breeds, including araucanas, heritage skyline and cream leg bar) eggs in the same box or carton, while the supermarkets at the same time sell mostly eggs from the larger producers, of the color preferred in that nation or region.

Very dark brown eggs of Marans, a French breed of chicken

These cultural trends have been observed for many years. *The New York Times* reported during the Second World War that housewives in Boston preferred brown eggs and those in New York preferred white eggs. In February 1976, the New Scientist magazine, in discussing issues of chicken egg color, stated "Housewives are particularly fussy about the colour of their eggs, preferring even to pay more for brown eggs although white eggs are just as good". As a result of these trends, brown eggs are usually more expensive to purchase in regions where white eggs are considered 'normal', due to lower production. In France and the United Kingdom, it is very difficult to buy white eggs, with most supermarkets supplying only the more popular brown eggs. By direct contrast, in Egypt it is very hard to source brown eggs, as demand is almost entirely for white ones, with the country's largest supplier describing white eggs as "table eggs" and packaging brown eggs for export.

Research conducted by a France institute in the 1970s demonstrated blue chicken eggs from the Chilean araucana fowl can be stronger and more resilient to breakage.

Research at Nihon University, Japan in 1990 revealed a number of different issues were important to Japanese housewives when deciding which eggs to buy; however, color was a distinct factor, with most Japanese housewives preferring the white color.

Egg producers carefully consider cultural issues, as well as commercial ones, when selecting the breed or breeds of chicken used for production, as egg color varies between breeds. Among producers and breeders, brown eggs are often referred to as "tinted", while the speckled eggs preferred by some consumers are often referred to as being "red" in color.

## Living Conditions of Birds

Commercial factory farming operations often involve raising the hens in small, crowded cages, preventing the chickens from engaging in natural behaviors, such as wing-flapping, dust-bathing, scratching, pecking, perching, and nest-building. Such restrictions can lead to pacing and escape behavior.

Many hens confined to battery cages, and some raised in cage-free conditions, are debeaked to prevent them from harming each other and cannibalism. According to critics of the practice, this can cause hens severe pain to the point where some may refuse to eat and starve to death. Some hens may be force molted to increase egg quality and production level after the molting. Molting can be induced by extended feed withdrawal, water withdrawal or controlled lighting programs.

Laying hens are often slaughtered between 100 and 130 weeks of age, when their egg productivity starts to decline. Due to modern selective breeding, laying hen strains differ from meat production strains. As male birds of the laying strain do not lay eggs and are not suitable for meat production, they are generally killed soon after they hatch.

Free-range eggs are considered by some advocates to be an acceptable substitute to factory-farmed eggs. Free-range laying hens are given outdoor access instead of being contained in crowded cages. Questions on the actual living conditions of free-range hens have been raised in the United States of America, as there is no legal definition or regulations for eggs labeled as free-range in that country.

In the United States, increased public concern for animal welfare has pushed various egg producers to promote eggs under a variety of standards. The most widespread standard in use is determined by United Egg Producers through their voluntary program of certification. The United Egg Producers program includes guidelines regarding housing, food, water, air, living space, beak trimming, molting, handling, and transportation, however, opponents such as The Humane Society have alleged UEP Certification is misleading and allows a significant amount of unchecked animal cruelty. Other standards include "Cage Free", "Natural", "Certified Humane", and "Certified Organic". Of these standards, "Certified Humane", which carries requirements for stocking density and cage-free keeping and so on, and "Certified Organic", which requires hens to have outdoor access and be fed only organic vegetarian feed and so on, are the most stringent.

Effective 1 January 2012, the European Union banned conventional battery cages for egg-laying hens, as outlined in EU Directive 1999/74/EC. The EU permits the use of "enriched" furnished cages that must meet certain space and amenity requirements. Egg producers in many member states have objected to the new quality standards while in some countries even furnished cages and family cages are subject to be banned as well. The production standard of the eggs is visible on the mandatory egg marking where the EU egg code begins with 3 for caged chicken to 1 for free-range eggs and 0 for organic egg production.

## Killing of Male Chicks

In battery cage and free-range egg production, unwanted male chicks are killed at birth during the process of securing a further generation of egg-laying hens.

## Cultural Significance

A popular Easter tradition in some parts of the world is a decoration of hard-boiled eggs (usually by dying, but often by spray-painting). Adults often hide the eggs for children to find, an activity known as an Easter egg hunt. A similar tradition of egg painting exists in areas of the world influenced by the culture of Persia. Before the spring equinox in the Persian New Year tradition (called *Norouz*), each family member decorates a hard-boiled egg and sets them together in a bowl.

The tradition of a dancing egg is held during the feast of Corpus Christi in Barcelona and other Catalan cities since the 16th century. It consists of an emptied egg, positioned over the water jet from a fountain, which starts turning without falling.

Although a food item, eggs are sometimes thrown at houses, cars, or people. This act, known commonly as "egging" in the various English-speaking countries, is a minor form of vandalism and, therefore, usually a criminal offense and is capable of damaging property (egg whites can degrade certain types of vehicle paint) as well as causing serious eye injury. On Halloween, for example, trick or treaters have been known to throw eggs (and sometimes flour) at property or people from whom they received nothing. Eggs are also often thrown in protests, as they are inexpensive and nonlethal, yet very messy when broken.

## Feather Meal

Feather meal is a byproduct of processing poultry; it is made from poultry feathers by partially grinding them under elevated heat and pressure, and then grinding and drying. Although total nitrogen levels are fairly high (up to 12%), the bioavailability of this nitrogen may be low. Feather meal is used in formulated animal feed and in organic fertilizer.

Worldwide, more than 25 billion chickens are used for human consumption. Feather meal is made through a process called rendering. Steam pressure cookers with temperatures over 140 °C are used to "cook" and sterilize the feathers. This partially hydrolyzes the proteins, which denatures them. It is then dried, cooled and ground into a powder for use as a nitrogen source for animal feed (mostly ruminants) or as an organic soil amendment.

Containing up to 12% nitrogen, it is a source of slow-release, organic, high-nitrogen fertilizer for organic gardens. It is not water-soluble and does not make a good liquid fertilizer. It can be used to:

- Increase green leaf growth
- Activate compost decomposition
- Improve soil structure

When adding it to a garden as a nitrogen source, it must be blended into the soil to start the decomposition to make the nitrogenous compounds available to the plants. As an organic garden fertilizer, it is not synthetic or petroleum-based.

### Issues

A 2012 study found that feather meal contributes to inorganic arsenic exposure to humans whether it be in the form of animal feed or organic fertilizers.

An analysis of feather meal across the United States also shows a variety of drugs that chickens are exposed to. Included are antibiotics such as fluoroquinolones which have been banned but are still included in the food supply. Other drugs include Prozac, antihistamines, fungicide, sex hormone norgestimate, and caffeine. The bioacculumative effects of these variety of pharmaceuticals are of concern to human health.

## Down Feather

Down feather

The down of birds is a layer of fine feathers found under the tougher exterior feathers. Very young birds are clad only in down. Powder down is a specialized type of down found only in a few groups of birds. Down is a fine thermal insulator and padding, used in goods such as jackets, bedding (duvets), pillows and sleeping bags. The discovery of feathers trapped in ancient amber suggests that some species of dinosaur may have possessed down-like feathers.

### Description and Etymology

Like many precocial hatchlings, domestic chickens are already covered with a downy coat of feathers when they hatch.

The word *down* comes from the Old Norse word *dúnn*, which had the same meaning as its modern equivalent. The down feather is considered to be the "simplest" of all feather types. It has a short or vestigial rachis (shaft), few barbs, and barbules that lack hooks. There are three types of down: natal down, body down and powder down. Natal down is the layer of down feathers that cover most birds at some point in their early development. Precocial nestlings are already covered with a layer of down when they hatch, while altricial nestlings develop their down layer within days or weeks of hatching. Megapode hatchlings are the sole exception; they are already covered with contour feathers when they hatch. Body down is a layer of small, fluffy feathers that lie underneath the outer contour feathers on a bird's body.

## Powder Down

"Feather dust" left on a window after a birdstrike

Powder down, or pulviplumes, is a special type of down that occurs in a few groups of apparently unrelated birds. In some species, the tips of the barbules on powder down feathers disintegrate, forming fine particles of keratin, which appear as a powder, or "feather dust", among the feathers. These feathers grow continuously and are not molted. In other species, powder grains come from cells that surround the barbules of growing feathers. These specialized feathers are typically scattered among ordinary down feathers, though in some species, they occur in clusters. All parrots have powder down, with some species (including the mealy parrot) producing copious amounts. It is also found in tinamous and herons. The dust produced from powder down feathers is a known allergen in humans.

## Function

Body down feathers, like these exposed on this adult male budgerigar's back, lie underneath the contour feathers and help to insulate birds against heat loss.

The loose structure of down feathers traps air, which helps to insulate the bird against heat loss and contributes to the buoyancy of waterbirds. Species that experience annual temperature fluctuations typically have more down feathers following their autumn moult. There is some evidence that down feathers may also help to decrease the incidence of nestling cannibalism among some colonially nesting species, as the stiffness of the feathers make the young more difficult to swallow. Pollutants can reduce the efficiency of these functions. When oiled, for example, down feathers mat and clump together, which breaks down the bird's insulation and allows water to reach the skin.

Female wildfowl use down feathers plucked from their own breasts to line their scrape nests. This process performs the dual function of helping to insulate the eggs and exposing the female's brood pouch—an area of bare skin, rich in blood vessels, which transmits heat very efficiently. Of the various items birds use to line their nests, down feathers provide the most effective insulation, though only when dry; wet down is the least effective insulator. Down may also help camouflage the eggs when the female is away from the nest, particularly as the birds often draw the feathers over their eggs before leaving.

Because a bird can eliminate heavy metals in its feathers and because feathers can be collected non-invasively and stored indefinitely, down feathers can be used to check for evidence of metal contamination in the bird's environment. Studies have shown a high level of correlation between the level of metal contamination in a bird's diet and the level found in its feathers, with the proportion of the chemicals found in its feathers remaining relatively constant (and relatively high for some metals).

## Mutation

Mutations in the genes that control the formation of down feathers have been recorded in a German White Leghorn flock. Although the elements of a normal down feather are present, a hyperkeratosis of the feather's horny sheath after 16–17 days of incubation results in the sheath not splitting as it should during the final stages of the feather's growth. Because of that abnormal splitting, the bird's down appears to be matted; chicks with this condition look bristly and singed and tend to be lighter in body weight than normal chicks are.

## Human Use

Down feathers were used by indigenous North Americans for religious ceremonies and as powerful symbols. In the stories of some cultures, the down feathers of an eagle were important gifts given by the bird to the story's hero. In the Ghost Dance, a religious movement that became particularly widespread among the Plains Indians, each dancer held a painted feather that was tipped with a down feather painted with another color; the feathers were generally those of a crow, which was sacred to the Ghost Dance, or of an eagle, which was sacred to all tribes. Zuni prayer sticks were also made using eagle down. While eagle feathers belonged to the Sun Priest, who planted them to the sun, other priests could use them if rain was needed, as the down is said to suggest "fleecy clouds that gather on the horizon before rain". The Hopi rubbed eagle down feathers over rattlesnakes being collected for their Snake Dances, in an effort to soothe and calm the reptiles.

A female common eider sits on her nest, surrounded by down feathers.

For centuries, humans across the globe have used down feathers for insulation. Russian documents from the 1600s list "bird down" among the goods sold to Dutch merchants, and communities in northern Norway began protecting the nests of eider ducks as early as 1890. Eiders are still "farmed" by people in Iceland, Scandinavia and Siberia. The birds are provided with nest sites and protected from predators, and down is collected intermittently during the nesting season without harming the nests or female ducks. The first collection is made roughly halfway through the incubation period, when some 0.75 oz (21 g) of high quality down is removed per nest. When the eggs have hatched and the young have left the area, the remaining down and breast feathers are gathered, typically resulting in another 0.75 oz (21 g) of lower quality feathers per nest. In general, 50–60 nests will produce about one kilo (2.2 pounds) of down feathers. This means that only a few thousand pounds of eider down is collected from wild nests each year.

In Iceland and Scandinavia, colonies of more than 5,000 birds sometimes develop in "farmed" areas, while in some protected areas of the Novaya Zemlya archipelago, nest densities exceed 13,000 per hectare (more than 5,260 per acre). On the other hand, the more nomadic indigenous peoples of Arctic Canada did not see down production as a reasonable source of income and tended to overexploit eider nests with "indifferent" collection of down feathers from the nests. Although the down feathers of various species of wildfowl, gulls and other seabirds have historically been used for insulation, most now come from domestic geese. Some 70 percent of the world's supply comes from China, typically from birds killed for their meat. Most of the rest comes from Europe and Canada, from birds harvested for meat or pâté.

In the United States, Federal Trade Commission regulations require that any product labeled "100% Down" must contain only down feathers, while products labeled simply

"Down" can contain a mixture of fiber and feathers. In addition, products labeled as "Goose Down" must contain at least 90% goose down, 10% goose feathers. Down insulation is rated by fill power, which is the number of cubic inches displaced by a given ounce of down ($in^3/oz$). To measure fill power, an ounce of down is placed into a graduated cylinder, and a small weight is dropped in on top of it; the volume below the weight indicates the fill power. Eider down has the highest fill power, at 1200. However, even down with a fill power as low as 550 still provides reasonably good insulation. Higher fill-power downs will thus insulate better than lower fill-power downs of the same weight. Insulation in most outdoor equipment ranges from about 400 to 900 $in^3/oz$ (230–520 $cm^3/g$). Down rated 500–650 $in^3/oz$ (290–375 $cm^3/g$) is warm enough and light enough for most conditions, and 800–900 $in^3/oz$ (460–520 $cm^3/g$) fill is used for very lightweight and/or very cold-weather gear.

Down is warm, lightweight and packable. If well cared for, it retains its loft up to three times longer than do most synthetics. However, when it is wet, the thermal properties of down are virtually eliminated. Down forms clumps if exposed to dampness or moisture, and will mildew if left damp. In addition, it will absorb and retain odors.

Down feathers tend to become more allergenic as they age, as they become contaminated with molds and dust mites. Pillows provide the most common source of exposure, though mattresses, comforters, outerwear and upholstery can also cause problems. A Finnish study has shown that true feather allergies are rare, with most issues caused by dust mites.

## Animal Cruelty

A percentage of the world's supply of down feathers has been plucked from live birds, a practice which is condemned as cruel by animal welfare groups. The precise percentage of down harvested in this manner is uncertain; while some references report that it is only a small fraction of the total (less than 1% in 2011), a 2009 Swedish documentary reported that it might be as much as 50–80% of the total supply, a figure supported by IKEA (a home furnishing chain) and an industry representative. The documentary also shows birds lying on the floor with large flesh wounds from the plucking, after which the wounds were stitched using a needle and thread without anesthetic. Although live-plucking is illegal in the United States and Europe, it is known to occur in two European countries (Poland and Hungary) and in China. Public sentiment against the practice has, in some countries, been strong. IKEA and Patagonia (a clothing manufacturer) have altered product lines to eliminate the use (or possible use) of live-plucked down.

## Non-avian Dinosaurs

Feathers found in amber in western France, Canada and northeastern China suggest that some non-avian dinosaurs may have had primitive, down-like feathers.

### Types of Down

- Duck down:
    - Eider down: This is considered to be the ultimate in duck down; it is collected from the breast feathers of the female eider duck and has the largest duck down clusters.
- Goose down:
    - Polish/Hungarian goose down: Widely considered to be the finest quality down clusters, they are larger than the norm and are mostly pure white in color.
- Feather-down combinations: Down compacts with age and is not appropriate for some applications; most down-filled pillow forms are stuffed with a combination of down and feathers. The presence of feathers adds weight, stability and bulk. Therefore, a mixture with more down will be lighter and fluffier than a pillow with a higher feather ratio.
- Marabou(t): Marabou (historically spelt with a T) is down from the marabou stork although turkey down is used as a substitute. It is widely used as a fashion trimming and in fly tying.

## Chicken as Food

Chicken is the most common type of poultry in the world.

In developed countries, chickens are usually subject to intensive farming methods.

### History

The modern chicken is a descendant of red junglefowl hybrids along with the grey junglefowl first raised thousands of years ago in the northern parts of the Indian subcontinent.

Chicken as a meat has been depicted in Babylonian carvings from around 600 BC. Chicken was one of the most common meats available in the Middle Ages. It was eaten over most of the Eastern hemisphere and a number of different kinds of chicken such as capons, pullets and hens were eaten. It was one of the basic ingredients in the so-called white dish, a stew usually consisting of chicken and fried onions cooked in milk and seasoned with spices and sugar.

In the 1800s chicken was more expensive than other meats and it was "sought by the rich because [it is] so costly as to be an uncommon dish." Chicken consumption in the United States increased during World War II due to a shortage of beef and pork. In

Europe, consumption of chicken overtook that of beef and veal in 1996, linked to consumer awareness of Bovine spongiform encephalopathy (mad cow disease).

Chicken fry

## Breeding

Modern varieties of chicken such as the Cornish Cross, are bred specifically for meat production, with an emphasis placed on the ratio of feed to meat produced by the animal. The most common breeds of chicken consumed in the US are Cornish and White Rock.

Chickens raised specifically for food are called broilers. In the United States, broilers are typically butchered at a young age. Modern Cornish Cross hybrids, for example, are butchered as early as 8 weeks for fryers and 12 weeks for roasting birds.

Capons (castrated cocks) produce more and fattier meat. For this reason, they are considered a delicacy and were particularly popular in the Middle Ages.

## Edible Components

Chicken in a public market

- Main
    - Breast: These are white meat and are relatively dry.

- Leg: Comprises two segments:
    1. The "drumstick"; this is dark meat and is the lower part of the leg,
    2. the "thigh"; also dark meat, this is the upper part of the leg.
- Wing: Often served as a light meal or bar food. Buffalo wings are a typical example. Comprises three segments:
    1. the "drumette", shaped like a small drumstick,
    2. the middle "flat" segment, containing two bones, and
    3. the tip, sometimes discarded.

- Other
    - Chicken feet: These contain relatively little meat, and are eaten mainly for the skin and cartilage. Although considered exotic in Western cuisine, the feet are common fare in other cuisines, especially in the Caribbean and China.
    - Giblets: organs such as the heart, gizzards, and liver may be included inside a butchered chicken or sold separately.
    - Head: Considered a delicacy in China, the head is split down the middle, and the brains and other tissue is eaten.
    - Kidneys: Normally left in when a broiler carcass is processed, they are found in deep pockets on each side of the vertebral column.
    - Neck: This is served in various Asian dishes. It is stuffed to make helzel among Ashkenazi Jews.
    - Oysters: Located on the back, near the thigh, these small, round pieces of dark meat are often considered to be a delicacy.
    - Pygostyle (chicken's buttocks) and testicles: These are commonly eaten in East Asia and some parts of South East Asia.
- By-products
    - Blood: Immediately after slaughter, blood may be drained into a receptacle, which is then used in various products. In many Asian countries, the blood is poured into low, cylindrical forms, and left to congeal into disc-like cakes for sale. These are commonly cut into cubes, and used in soup dishes.

- Carcase and back: After the removal of the flesh, this is used for soup stock.
- Chicken eggs: The most well-known and well-consumed byproduct.
- Heart and gizzard: in Brazilian *churrascos*, chicken hearts are an often seen delicacy.
- Liver: This is the largest organ of the chicken, and is used in such dishes as Pâté and chopped liver.
- Schmaltz: This is produced by rendering the fat, and is used in various dishes.

## Health

Chicken meat contains about two to three times as much polyunsaturated fat than most types of red meat when measured as weight percentage.

Chicken generally includes low fat in the meat itself (castrated roosters excluded). The fat is highly concentrated on the skin. A 100g serving of baked chicken breast contains 4 grams of fat and 31 grams of protein, compared to 10 grams of fat and 27 grams of protein for the same portion of broiled, lean skirt steak.

However, according to a 2006 Harvard School of Public Health study of 135,000 people, people who ate grilled skinless chicken 5 or more times a week had a 52 percent higher chance of developing bladder cancer compared to people who did not. However, such strong associations were not found in individuals regularly consuming chicken with skin intact.

## Use of Roxarsone in Chicken Production

In many factory farms, chickens are routinely administered with the feed additive Roxarsone, an organoarsenic compound which partially decomposes into inorganic arsenic compounded in the flesh of chickens, and in their feces, which are often used as a fertilizer. The compound is used to control stomach pathogens and promote growth. In a 2013 sample conducted by the Johns Hopkins School of Public Health of chicken meat from poultry producers that did not prohibit roxarsone, 70% of the samples in the US had levels which exceeded the safety limits as set by the FDA. The FDA has since revised its stance on safe limits to inorganic arsenic in animal feed by stating that "any new animal drug that contributes to the overall inorganic arsenic burden is of potential concern."

## Antibiotic Resistance

Information obtained by the Canadian Integrated Program for Antimicrobial Resistance (CIPARS) "strongly indicates that cephalosporin resistance in humans is moving in lockstep with the use of the drug in poultry production." According to the Canadian Medical Association Journal, the unapproved antibiotic ceftiofur is routinely injected into eggs in Quebec and Ontario to discourage infection of hatchlings. Although the data are contested by the industry, antibiotic resistance in humans appears to be directly related to the antibiotic's use in eggs.

A recent study by the Translational Genomics Research Institute showed that nearly half (47%) of the meat and poultry in US grocery stores was contaminated with *S. aureus*, with more than half (52%) of those bacteria resistant to antibiotics. Furthermore, as per the FDA, more than 25% of retail chicken is resistant to 5 or more different classes of antibiotic treatment drugs in the United States. An estimated 90-100% of conventional chicken contains, at least, one form of antibiotic resistance microorganism. While organic chicken has been found to have a lower incidence at 84%.

## Fecal Matter Contamination

In random surveys of chicken products across the United States in 2012, the Physicians Committee for Responsible Medicine found 48% of samples to contain fecal matter.

On most commercial chicken farms (Concentrated Animal Feeding Operations) CAFO, the chickens spend their entire life standing in, laying on, and living in their own manure, which is somewhat mixed in with the bedding material (e.g. sawdust, wood shavings, chopped straw, etc.).

During shipping from the CAFO farm to the abattoir, the chickens are usually placed inside shipping crates that usually have slatted floors. Those crates are then piled 5 to 10 rows high on the transport truck to the abattoir. During shipment, the chickens tend to defecate, and that chicken manure tends to sit inside the crowded cages, contaminating the feathers and skin of the chickens, or rains down upon the chickens and crates on the lower levels of the transport truck. By the time the truck gets to the abattoir, most chickens have had their skin and feathers contaminated with feces.

There is also fecal matter in the intestines. While the slaughter process removes the feathers and intestines, only visible fecal matter is removed. The high speed automated processes at the abattoir are not designed to remove this fecal contamination on the feather and skin. The high speed processing equipment tend to spray the contamination around to the birds going down the processing line, and the equipment on the line itself. At one or more points on most abattoirs, chemical sprays and baths (e.g. bleach, acids, peroxides, etc.) are used to partially rinse off or kill this bacterial contamination. Unfortunately, the fecal contamination, once it has occurred, especially in the various membranes between the skin and muscle, is impossible to completely remove.

With slaughter lines processing up to 140 birds/minute, safety inspectors do not have adequate time to properly examine visible fecal matter. The USDA is currently allowing some abattoirs to process at unlimited line speeds (i.e. in excess of 140 birds/minute), further exasperating the fecal contamination issue.

## Marketing and Sales

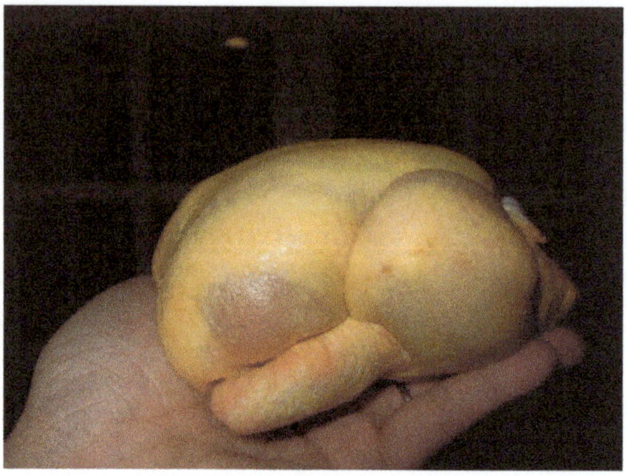

A poussin, or juvenile chicken

Juvenile chickens, of less than 28 days of age at slaughter in the United Kingdom are marketed as poussin. Mature chicken is sold as small, medium or large.

Whole mature chickens are marketed in the United States as fryers, broilers, and roasters. Fryers are the smallest size (2.5-4 lbs dressed for sale), and the most common, as chicken reach this size quickly (about 7 weeks). Most dismembered packaged chicken would be sold whole as fryers. Broilers are larger than fryers. They are typically sold whole. Roasters, or roasting hens, are the largest chickens commonly sold (3–5 months and 6-8 lbs) and are typically more expensive. Even larger and older chickens are called stewing chickens but these are no longer usually found commercially. The names reflect the most appropriate cooking method for the surface area to volume ratio. As the size increases, the volume (which determines how much heat must enter the bird for it to be cooked) increases faster than the surface area (which determines how fast heat can enter the bird). For a fast method of cooking, such as frying, a small bird is appropriate: frying a large piece of chicken results in the inside being undercooked when the outside is ready.

Chicken is also sold in dismembered pieces. Pieces may include quarters, or fourths of the chicken. A chicken is typically cut into two leg quarters and two breast quarters. Each quarter contains two of the commonly available pieces of chicken. A leg quarter contains the thigh, drumstick and a portion of the back; a leg has the back portion removed. A breast quarter contains the breast, wing and portion of the back; a breast has the back portion and wing removed. Pieces may be sold in packages of all of the same pieces, or in combination packages. Whole chicken cut up refers to either the entire bird cut into 8 individual pieces. (8-piece cut); or sometimes without the back. A 9-piece cut (usually for fast food restaurants) has the tip of the breast cut off before splitting. Pick of the chicken, or similar titles, refers to a package with only some of the chicken pieces. Typically the breasts, thighs, and legs without wings or back. Thighs and breasts are sold boneless and/or skinless. Dark meat (legs, drumsticks and thighs) pieces are typically cheaper than white meat pieces (breast, wings). Chicken livers and/or gizzards are commonly available packaged separately. Other parts of the chicken, such as the neck, feet, combs, etc. are not widely available except in countries where they are in demand, or in cities that cater to ethnic groups who favor these parts.

There are many fast food restaurant chains on both a national and global scale that sell exclusively or primarily in poultry products including KFC (global), Red Rooster (Australia), Hector Chicken (Belgium) and CFC (Indonesia). Most of the products on the menu in such eateries are fried or breaded and are served with french fries.

## Cooking

Marination of chicken for grilling

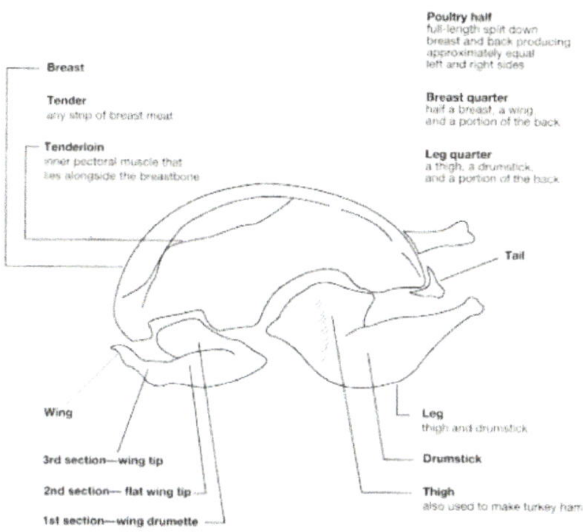


The United States Department of Agriculture classifies cuts of poultry in a manner similar to beef.

Chicken with mushrooms, tomatoes and spices.

Oven roasted chicken with potatoes.

Chicken Peking (Philippines)

Butter chicken masala (India)

Raw chicken may contain *salmonella*. The safe minimum cooking temperature recommended by the U.S. Department of Health & Human Services is 165 °F (74 °C) to prevent foodborne illness because of bacteria and parasites. However, in Japan raw chicken is sometimes consumed in a dish called *torisashi*, which is sliced raw chicken served in sashimi style. Another preparation is *toriwasa* which is lightly seared on the outsides while the inside remains raw.

Chicken can be cooked in many ways. It can be made into sausages, skewered, put in salads, grilled, breaded and deep-fried, or used in various curries. There is significant variation in cooking methods amongst cultures. Historically common methods include roasting, baking, broasting, and frying. Western cuisine frequently has chicken prepared by deep frying for fast foods such as fried chicken, chicken nuggets, chicken lollipops or buffalo wings. They are also often grilled for salads or tacos.

Chickens often come with labels such as "roaster", which suggest a method of cooking based on the type of chicken. While these labels are only suggestions, ones labeled for stew often do not do well when cooked with other methods.

Some chicken breast cuts and processed chicken breast products include the moniker "with rib meat". This is a misnomer, as it is the small piece of white meat that overlays the scapula, and is removed with the breast meat. The breast is cut from the chicken and sold as a solid cut, while the leftover breast and true rib meat is stripped from the bone through mechanical separation for use in chicken franks, for example. Breast meat is often sliced thinly and marketed as chicken slices, an easy filling for sandwiches. Often, the tenderloin (pectoralis minor) is marketed separately from the breast (pectoralis major). In the US, "tenders" can be either tenderloins or strips cut from the breast. In the UK the strips of pectoralis minor are called "chicken mini-fillets".

Chicken bones are hazardous to health as they tend to break into sharp splinters when eaten, but they can be simmered with vegetables and herbs for hours or even days to make chicken stock.

In Asian countries it is possible to buy bones alone as they are very popular for making chicken soups, which are said to be healthy. In Australia the rib cages and backs of chickens after the other cuts have been removed are frequently sold cheaply in supermarket delicatessen sections as either "chicken frames" or "chicken carcasses" and are purchased for soup or stock purposes.

## Freezing

Raw chicken maintains its quality longer in the freezer as compared to when having been cooked because moisture is lost during cooking. There is little change in nutrient value of chicken during freezer storage. For optimal quality, however, a maximal storage time in the freezer of 12 months is recommended for uncooked whole chicken, 9 months for uncooked chicken parts, 3 to 4 months for uncooked chicken giblets, and 4 months for cooked chicken. Freezing doesn't usually cause color changes in poultry, but the bones and the meat near them can become dark. This bone darkening results when pigment seeps through the porous bones of young poultry into the surrounding tissues when the poultry meat is frozen and thawed.

It is safe to freeze chicken directly in its original packaging, however this type of wrap is permeable to air and quality may diminish over time. Therefore, for prolonged storage, it is recommended to overwrap these packages. It is recommended to freeze unopened vacuum packages as is. If a package has accidentally been torn or has opened while food is in the freezer, the food is still safe to use, but it is still recommended to overwrap or rewrap it. Chicken should be away from other foods, so if they begin to thaw, their juices won't drip onto other foods. If previously frozen chicken is purchased at a retail store, it can be refrozen if it has been handled properly.

Chicken can be cooked or reheated from the frozen state, but it will take approximately one and a half times as long to cook, and any wrapping or absorbent paper should be discarded. There are three generally accepted safe methods of reheating frozen chicken: in the refrigerator, in cold water, or using a microwave oven. These methods are endorsed by the FDA as safe, as they minimize the risk of bacterial growth. Bacteria survives but does not grow in freezing temperatures. However, if frozen cooked foods are not defrosted properly and are not reheated to temperatures that kill bacteria, chances of getting a foodborne illness greatly increase.

## Duck as Food

Duck breast with foie gras

Braised duck, Teochew style

Cantonese style roast duck with rice

Duck roasted with Chinese angelica herb

In food terminology, duck or duckling (when meat comes from a juvenile duck) refers to duck meat, the meat of several species of bird in the family Anatidae, found in both fresh and salt water.

One species of freshwater duck, the mallard, has been domesticated and is a common livestock bird in many cultures. Duck is eaten in various cuisines around the world.

Magret refers specifically to the breast of a mulard or Muscovy (or Barbary) duck that has been force fed to produce foie gras.

## Duck Meat

Duck meat is derived primarily from the breasts and legs of ducks. The meat of the legs is darker and somewhat fattier than the meat of the breasts, although the breast meat is darker than the breast meat of a chicken or a turkey. Being waterfowl, ducks have a layer of heat-insulating subcutaneous fat between the skin and the meat.

De-boned duck breast can be grilled like steak, usually leaving the skin and fat on.

Internal organs such as heart and kidneys may also be eaten; the liver in particular is often used as a substitute for goose liver in foie gras.

## Types of Duck Meat in the US

The most common duck meat consumed in the United States is the Pekin duck. Because most commercially raised Pekins come from Long Island, New York, Pekins are also sometimes called "Long Island" ducks, despite being of Chinese origin. Some specialty breeds have become more popular in recent years, notably the Muscovy duck, and the mulard duck (a sterile hybrid of Pekins and Muscovies). Unlike most other domesticated ducks, Muscovy ducks are not descended from mallards.

Duck is particularly predominant in the Chinese cuisine — a popular dish is Peking duck, which is made from the Pekin duck. They are commonly eaten with scallions, cucumbers and hoisin sauce wrapped in a tortilla-like pancake made of flour and water or a soft, risen bun known as gebao (割包). According to the USDA, nearly 26 million ducks were eaten in the U.S. in 2004.

## Duck in Dishes from Around the World

Duck is used in a variety of dishes around the world, most of which involve roasting for at least part of the cooking process to aid in crisping the skin. Notable duck dishes include:

- Bebek Betutu – a famous traditional dish from Bali, Indonesia. The duck is first seasoned with pungent roots and various herbs, wrapped with banana leaves, and roasted. Chicken is also used to prepare Betutu.
- Confit – duck legs that have been cured (partly or fully) in salt, then marinated and poached in duck fat, typically with garlic and other herbs. The French word *confit* means "preserved", and the French name for duck confit is "confit de canard".
- Czernina – a sweet and sour Polish soup made of duck blood and clear poultry broth. It was once considered a symbol of Polish culture until the 19th century, customarily served to young men and is even featured as a plot device in a famous epic poem called Pan Tadeusz.
- Duck à l'orange – a classic French dish in which the duck is roasted and served with an orange sauce.
- Foie gras – a specially fattened and rich liver, or a pâté made from the liver, sometimes taken from a duck but usually from a goose.
- Long Island roast duckling – this is a whole roasted bird, sometimes brined previously. When done properly, most of the fat melts off during the cooking process, leaving a crispy skin and well-done meat. Some restaurants on Long Island serve this dish with a cherry sauce.
- Oritang – a variety of *guk*, Korean soup made with duck and various vegetables.
- Peking duck – a famous Chinese dish originating from Beijing, prepared since the Ming Dynasty era. It is prized for the thin, crispy skin, with authentic versions of the dish serving mostly the skin and little meat, and eaten with pan-

cakes, scallions, and hoisin sauce or sweet bean sauce.

- Pressed duck – a complex dish originally from Rouen, France.
- Turducken: an American dish that comprises a turkey, stuffed with a duck, which is in turn stuffed with a chicken.
- Zhangcha duck – a quintessential dish of Sichuan cuisine. It is first prepared by smoking a marinated duck over tea leaves and twigs of the camphor plant, then steamed, and finally deep fried for a crisp finish. Also called tea-smoked duck.

## Balut

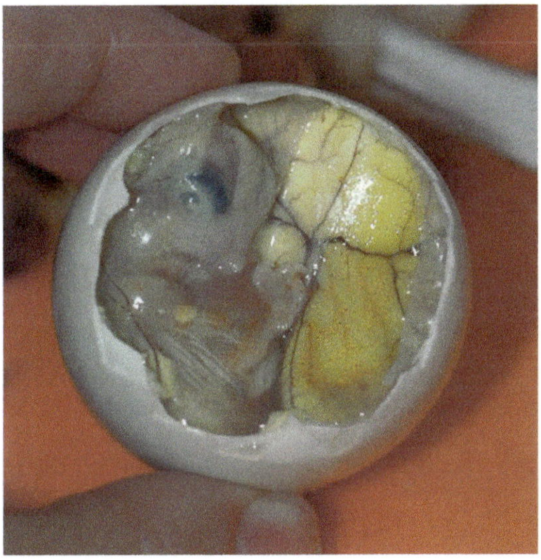

Partially shelled balut egg showing yolk and embryo.

A balut is a developing bird embryo (usually a duck or chicken) that is boiled and eaten from the shell. It originates and is commonly sold as street-food in the Philippines. They are common food in countries in Southeast Asia, such as Laos (*khai look* ໄຂ່ລູກ in Lao), Cambodia (*pong tia koun* ពងទាកូន in Cambodian), Thailand (*Khai Khao* ไข่ข้าว in Thai) and Vietnam (*trứng vịt lộn* or *hột vịt lộn* in Vietnamese). They are often served with beer. The Tagalog and Malay word *balut* means “wrapped”.

The length of incubation before the egg is cooked is a matter of local preference, but generally ranges between 14 and 21 days.

## Pollution Contaminating Wild Duck

Ducks caught in the wild may be contaminated from pollution of rivers and other bodies of water, because they eat fish and other aquatic life. In particular, PCBs may pose a health risk for those who eat wild duck frequently.

## References


- Brothwell, Don R.; Patricia Brothwell (1997). Food in Antiquity: A Survey of the Diet of Early Peoples. Johns Hopkins University Press. pp. 54–55. ISBN 0-8018-5740-6.
- McGee, H. (2004). On Food and Cooking: The Science and Lore of the Kitchen. New York: Scribner. ISBN 0-684-80001-2.
- This was experimentally shown and documented in the book The Food Lab: Better Home Cooking Through Science by J. Kenji López-Alt; published September 21, 2015 by W. W. Norton & Company, ISBN 978-0393081084
- Spearman, Richard Ian Campbell (1973). The Integumen: A Textbook of Skin Biology. Cambridge University Press. p. 97. ISBN 0-521-20048-2.
- Juniper, Tony; Parr, Mike (2003). Parrots: A Guide to the Parrots of the World. London: Christopher Helm. p. 17. ISBN 0-7136-6933-0.
- Kalstone, Shirlee (2006). Allergic to Pets?: The Breakthrough Guide to Living with the Animals You Love. New York, NY: Bantam Dell. pp. 34–35. ISBN 0-553-38367-1.
- Carboneras, Carles (1992). "Family Anatidae (Ducks, Geese and Swans)". Handbook of Birds of the World, Volume 1: Ostrich to Ducks. Barcelona, Spain: Lynx Edicions. p. 558. ISBN 84-87334-10-5.
- Mooney, James (1973). The Ghost Dance Religion and Wounded Knee. Mineola, NY: Courier Dover Publications. p. 919. ISBN 0-486-26759-8.
- Johnsgard, Paul A. (1999). Earth, Water and Sky: A Naturalist's Stories and Sketches. Austin, TX: University of Texas Press. pp. 119–120. ISBN 0-292-74059-X.
- Roth, Harald H.; Merz, Günter (1997). Wildlife Resources: A Global Account of Economic Use. Berlin: Springer-Verlag. p. 363. ISBN 3-540-61357-9.
- Randall, Glenn (2000). The Outward Bound Staying Warm in the Outdoors Handbook. New York, NY: Lyons Press. p. 63. ISBN 1-58574-089-6.
- Wilks, Brian B. (2004). Browsing Science Research at the Federal Level in Canada: History, Research. Toronto, ON: University of Toronto Press. p. 210. ISBN 0-8020-8811-2.
- Cooper, Donald C. (2005). Fundamentals of Search and Rescue. Sudbury, MA: Jones and Bartlett Learning. p. 89. ISBN 0-7637-4807-2.
- Hellekson, Terry (2005). Fish flies : the encyclopedia of the fly tier's art (1st ed.). Salt Lake City: Gibbs Smith. p. 91. ISBN 9781586856922.

# 7

# Poultry Diseases

Aspergillosis is a group of diseases which is caused by a fungus named aspergillus. It causes diseases such as tuberculosis or chronic obstructive pulmonary disease. A number of diseases have been mentioned in this chapter, some of these are avian infectious bronchitis, botulism, infectious bursal disease and marek's disease. The subject matter of this text is very important as poultry can cause disease that can cause death and infections.

## Aspergillosis

Aspergillosis is the name given to a wide variety of diseases caused by infection by fungi of the genus Aspergillus. The majority of cases occur in people with underlying illnesses such as tuberculosis or chronic obstructive pulmonary disease (COPD), but with otherwise healthy immune systems. Most commonly, aspergillosis occurs in the form of chronic pulmonary aspergillosis (CPA), aspergilloma or allergic bronchopulmonary aspergillosis (ABPA). Some forms are intertwined; for example ABPA and simple aspergilloma can progress to CPA.

Other, non-invasive manifestations include fungal sinusitis (both allergic in nature and with established fungal balls), otomycosis (ear infection), keratitis (eye infection) and onychomycosis (nail infection). In most instances these are less severe, and curable with effective antifungal treatment.

People with deficient immune systems—such as patients undergoing hematopoietic stem cell transplantation, chemotherapy for leukaemia, or AIDS—are at risk of more disseminated disease. Acute invasive aspergillosis occurs when the immune system fails to prevent Aspergillus spores from entering the bloodstream via the lungs. Without the body mounting an effective immune response, fungal cells are free to disseminate throughout the body and can infect major organs such as the heart and kidneys.

The most frequently identified pathogen is *Aspergillus fumigatus*—a ubiquitous organism that is capable of living under extensive environmental stress. It is estimated that most humans inhale thousands of Aspergillus spores daily, but they do not affect most people's health due to effective immune responses. Taken together, the major chronic, invasive and allergic forms of aspergillosis account for around 600,000 deaths annually worldwide.

### Symptoms

A fungus ball in the lungs may cause no symptoms and may be discovered only with a chest X-ray, or it may cause repeated coughing up of blood, chest pain, and occasionally severe, even fatal, bleeding. A rapidly invasive *Aspergillus* infection in the lungs often causes cough, fever, chest pain, and difficulty breathing.

Poorly controlled aspergillosis can disseminate through the blood stream to cause widespread organ damage. Symptoms include fever, chills, shock, delirium, seizures and blood clots. The person may develop kidney failure, liver failure (causing jaundice), and breathing difficulties. Death can occur quickly.

Aspergillosis of the ear canal causes itching and occasionally pain. Fluid draining overnight from the ear may leave a stain on the pillow. Aspergillosis of the sinuses causes a feeling of congestion and sometimes pain or discharge. It can extend beyond the sinuses.

In addition to the symptoms, an X-ray or computerised tomography (CT) scan of the infected area provides clues for making the diagnosis. Whenever possible, a doctor sends a sample of infected material to a laboratory to confirm identification of the fungus.

## Diagnosis

On chest X-ray and CT, pulmonary aspergillosis classically manifests as a halo sign, and, later, an air crescent sign. In hematologic patients with invasive aspergillosis, the galactomannan test can make the diagnosis in a noninvasive way. False positive Aspergillus galactomannan tests have been found in patients on intravenous treatment with some antibiotics or fluids containing gluconate or citric acid such as some transfusion platelets, parenteral nutrition or PlasmaLyte.

On microscopy, *Aspergillus* species are reliably demonstrated by silver stains, e.g., Gridley stain or Gomori methenamine-silver. These give the fungal walls a gray-black colour. The hyphae of Aspergillus species range in diameter from 2.5 to 4.5 µm. They have septate hyphae, but these are not always apparent, and in such cases they may be mistaken for Zygomycota. *Aspergillus* hyphae tend to have dichotomous branching that is progressive and primarily at acute angles of about 45°.

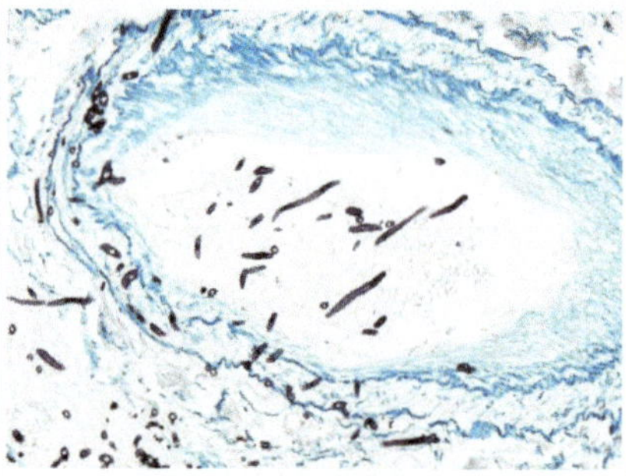

Angio invasive pulmonary aspergillosis

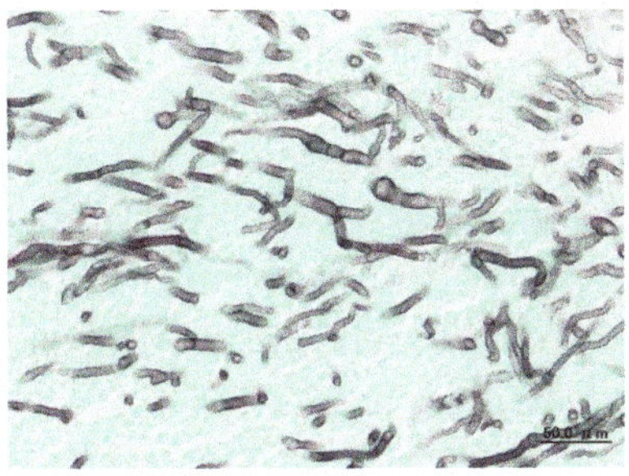

Angio invasive pulmonary aspergillosis (closeup)

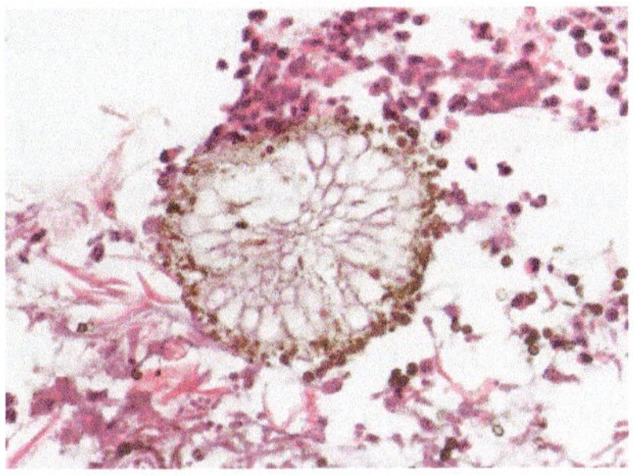

Aspergillus vesicle (HE stain).

## Treatment

The current medical treatments for aggressive invasive Aspergillosis include voriconazole and liposomal amphotericin B in combination with surgical debridement. For the less aggressive allergic bronchopulmonary aspergillosis findings suggest the use of oral steroids for a prolonged period of time, preferably for 6–9 months in allergic aspergillosis of the lungs. Itraconazole is given with the steroids, as it is considered to have a "steroid sparing" effect, causing the steroids to be more effective, allowing a lower dose.,

Other drugs used, such as amphotericin B, caspofungin (in combination therapy only), flucytosine (in combination therapy only), or itraconazole, are used to treat this fungal infection. However, a growing proportion of infections are resistant to the triazoles. *A. fumigatus*, the most commonly infecting species, is intrinsically resistant to fluconazole.

## Prevention

Prevention of Aspergillosis involves a reduction of mold exposure via environmental infection-control. Anti-fungal prophylaxis can be given to high-risk patients. Posaconazole is often given as prophylaxis in severely immunocompromised patients.

### Infections in Animals

Albeit relatively rare in humans, aspergillosis is a common and dangerous infection in birds, particularly in pet parrots. Mallards and other ducks are particularly susceptible, as they will often resort to poor food sources during bad weather. Captive raptors, such as falcons and hawks, are susceptible to this disease if they are kept in poor conditions and especially if they are fed pigeons, which are often carriers of "asper". It can be acute in chicks, but chronic in mature birds.

Aspergillosis has been the culprit in several rapid die-offs among waterfowl. From 8 December until 14 December 2006, over 2,000 Mallards died in the Burley, Idaho area of the USA, an agricultural community approximately 150 miles southeast of Boise. Mouldy waste grain from the farmland and feedlots in the area is the suspected source. A similar aspergillosis outbreak caused by mouldy grain killed 500 Mallards in Iowa, USA, in 2005.

While there is no connection between aspergillosis and the H5N1 strain of avian influenza (commonly called "bird flu"), rapid die-offs caused by aspergillosis can spark fears of bird flu outbreaks. Laboratory analysis is the only way to distinguish bird flu from aspergillosis.

In dogs, aspergillosis is an uncommon disease typically affecting only the nasal passages (nasal aspergillosis). This is much more common in dolicocephalic breeds. It can also spread to the rest of the body; this is termed disseminated aspergillosis and is rare, usually affecting individuals with underlying immune disorders.

## Avian Infectious Bronchitis

Avian infectious bronchitis (IB) is an acute and highly contagious respiratory disease of chickens. The disease is caused by avian infectious bronchitis virus (IBV), a coronavirus, and characterized by respiratory signs including gasping, coughing, sneezing, tracheal rales, and nasal discharge. In young chickens, severe respiratory distress may occur. In layers, respiratory distress, nephritis, decrease in egg production, and loss of internal (watery egg white) and external (fragile, soft, irregular or rough shells, shell-less) egg quality are reported.

### Etiology

IBV was the first coronavirus described and varies greatly genetically and phenotypically, with hundreads of serotypes and strains described. Coronaviruses contain the largest known viral RNA genome in number of nucleotides, of approximately 30,000 bases. The RNA forms a single strand and single segment. IBV diversity is based on transcriptional error, which may become very relevant if occurring in genomic sequences coding for proteins, involved in adsorption to target cell or inducing immune responses. Transcriptional error variants may emerge with evolutional advantage in susceptible chickens. Large genomic changes will occur with entire gene interchanges, by reassortment, as for its replication, seven subgenomic mRNAs are produced and will enable reassortment in coinfections.

## Clinical Signs

Coughing and rattling are common, most severe in young, such as broilers, and rapidly spreading in chickens confined or at proximity. Morbidity is 100% in non-vaccinated flocks. Mortality varies according to the virus strain (up to 60% in non-vaccinated flocks). Respiratory signs will subdue within two weeks. However, for some strains, a kidney infection may follow, causing mortality by toxemia. Younger chickens may die of tracheal occlusion by mucus (lower end) or by kidney failure. The infection may prolong in the cecal tonsils.

In laying hens, there can be transient respiratory signs, but mortality may be negligible. However, egg production drops sharply. A great percentage of produced eggs are misshapen and discolored. Many laid eggs have a thin or soft shell and poor albumen (watery), and are not marketable or proper for incubation. Normally-colored eggs, indicative of normal shells for instance in brown chickens, have a normal hatchability.

Egg yield curve may never return to normal. Milder strains may allow normal production after around eight weeks.

Read more about clinical AIB here

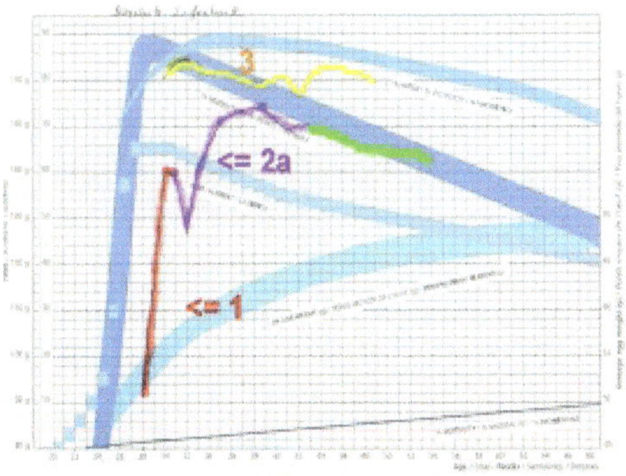

Egg yield curve in BI in a parent flock

Thin-shelled egg

Abnormal granulations on shell

Soft-shelled eggs

Misshapen and discolored eggs

## Diagnosis

Chicken respiratory diseases are difficult to differentiate and may not be diagnosed based on respiratory signs and lesions. Other diseases such as mycoplasmosis by Mycoplasma gallisepticum (chronic respiratory disease), Newcastle disease by mesogenic strains of Newcastle diseases virus (APMV-1), avian metapneumovirus, infectious laryngotracheitis, avian infectious coryza in some stages may clinically resemble IB. Similar kidney lesions may be caused by different etiologies, including other viruses, such as infectious bursal disease virus (the cause of Gumboro disease) and toxins (for instance ochratoxins of Aspergillus ochraceus), and dehydration.

In laying hens, abnormal and reduced egg production are also observed in Egg Drop Syndrome 76 (EDS), cause by an Atadenovirus and avian metapneumovirus infections. At present, IB is more common and far more spread than EDS. The large genetic and phenotypic diversity of IBV have been resulting in common vaccination failures. In addition, new strains of IBV, not present in commercial vaccines, can cause the disease in IB vaccinated flocks. Attenuated vaccines will revert to virulence by consecutive passage in chickens in densely populated areas, and may reassort with field strains, generating potentially important variants.

Definitive diagnosis relies on viral isolation and characterization. For virus characterization, recent methodology using genomic amplification (PCR) and sequencing of products, will enable very precise description of strains, according to the oligonucleotide primers designed and target gene. Methods for IBV antigens detection may employ labelled antibodies, such as direct immunofluorescence or immunoperoxidase. Antibodies to IBV may be detected by indirect immunofluorescent antibody test, ELISA and Haemagglutination inhibition (haemagglutinating IBV produced after enzymatic treatment by phospholipase C).

### Treatment and Prevention

No specific treatment is available, but antibiotics can be used to prevent secondary infections.

Vaccines are available (ATCvet codes: QI01AA03 (WHO) for the inactivated vaccine, QI01AD07 (WHO) for the live vaccine; plus various combinations).

Biosecurity protocols including adequate isolation, disinfection are important in controlling the spread of the disease.

## Avian Influenza

Avian influenza — known informally as *avian flu* or bird flu — refers to "influenza caused by viruses adapted to birds." The type with the greatest risk is highly pathogenic avian influenza (HPAI).

"Bird flu" is a phrase similar to "swine flu," "dog flu," "horse flu," or "human flu" in that it refers to an illness caused by any of many different strains of influenza viruses that have adapted to a specific host. Out of the three types of influenza viruses (A, B, and C), influenza A virus is a zoonotic infection with a natural reservoir almost entirely in birds. Avian influenza, for most purposes, refers to the influenza A virus.

Though influenza A is adapted to birds, it can also stably adapt and sustain person-to person transmission. Recent influenza research into the genes of the Spanish flu virus shows it to have genes adapted from both human and avian strains. Pigs can also be infected with human, avian, and swine influenza viruses, allow for mixtures of genes (reassortment) to create a new virus, which can cause an antigenic shift to a new influenza A virus subtype which most people have little to no immune protection.

Avian influenza strains are divided into two types based on their pathogenicity: high pathogenicity (HP) or low pathogenicity (LP). The most well-known HPAI strain, H5N1, appeared in China in 1996, and also has low pathogenic strains found in North America. Companion birds in captivity are unlikely to contract the virus and there has been no report of a companion bird with avian influenza since 2003. Pigeons do not contract or spread the virus.

## History

The most widely quoted date for the beginning of recorded history of avian influenza (initially known as fowl plague) was in 1878 when it was differentiated from other diseases that caused high mortality rates in birds. Fowl plague, however, also included Newcastle disease until as recently as the 1950s. Between 1959 and 1995, there were 15 recorded occasions of the emergence of HPAI viruses in poultry, but losses were minimal. Between 1996 and 2008 however, HPAI outbreaks in poultry have occurred at least 11 times and 4 of these outbreaks have involved millions of birds.

In the 1990s, the world's poultry population grew 76% in developing countries and 23% in developed countries, contributing to the increased prevalence of avian influenza. Before the 1990s, HPAI caused high mortality in poultry, but infections were sporadic and contained. Outbreaks have become more common due to the high density and frequent movement of flocks from intensive poultry production.

Influenza A/H5N1 was first isolated from a goose in China in 1996. Human infections were first reported in 1997 in Hong Kong. Since 2003, more than 700 human cases of Asian HPAI H5N1 have been reported to the WHO, primarily from 15 countries in Asia, Africa, the Pacific, Europe, and the Middle East, though over 60 countries have been affected.

## Genetics

Genetic factors in distinguishing between "human flu viruses" and "avian flu viruses" include:

> PB2: (RNA polymerase): Amino acid (or residue) position 627 in the PB2 protein encoded by the PB2 RNA gene. Until H5N1, all known avian influenza viruses had a Glu at position 627, while all human influenza viruses had a Lys.
>
> HA: (hemagglutinin): Avian influenza HA viruses bind alpha 2-3 sialic acid receptors, while human influenza HA viruses bind alpha 2-6 sialic acid receptors. Swine influenza viruses have the ability to bind both types of sialic acid receptors. Hemagglutinin is the major antigen of the virus against which neutralizing antibodies are produced, and influenza virus epidemics are associated with changes in its antigenic structure. This was originally derived from pigs, and should technically be referred to as "pig flu".

## Subtypes

There are many subtypes of avian influenza viruses, but only some strains of five subtypes have been known to infect humans: H5N1, H7N3, H7N7, H7N9, and H9N2. At least one person, an elderly woman in Jiangxi Province, China, died of pneumonia in December 2013 from the H10N8 strain, the first human fatality confirmed to be caused by that strain.

Most human cases of the avian flu are a result of either handling dead infected birds or from contact with infected fluids. It can also be spread through contaminated surfaces and droppings. While most wild birds mainly have only a mild form of the H5N1 strain, once domesticated birds such as chickens or turkeys are infected, it can become much more deadly because the birds are often in close contact. There is currently a large threat of this in Asia with infected poultry due to low hygiene conditions and close quarters. Although it is easy for humans to contract the infection from birds, human-to-human transmission is more difficult without prolonged contact. However, public health officials are concerned that strains of avian flu may mutate to become easily transmissible between humans. Some strains of avian influenza are present in the intestinal tract of large numbers of shore birds and water birds, but these strains rarely cause infection.

Spreading of H5N1 from Asia to Europe is much more likely caused by both legal and illegal poultry trades than dispersing through wild bird migrations, being that in recent studies, there were no secondary rises in infection in Asia when wild birds migrate south again from their breeding grounds. Instead, the infection patterns followed transportation such as railroads, roads, and country borders, suggesting poultry trade as being much more likely. While there have been strains of avian flu to exist in the United States, such as Texas in 2004, they have been extinguished and have not been known to infect humans.

Highly pathogenic strains spread quickly among flocks and can destroy a flock within 28 hours; the less pathogenic strains may affect egg production but are much less deadly.

## Examples of Avian Influenza A Virus Strains:

| HA subtype designation | NA subtype designation | Avian influenza A viruses |
|---|---|---|
| H1 | N1 | A/duck/Alberta/35/76(H1N1) |
| H1 | N8 | A/duck/Alberta/97/77(H1N8) |
| H2 | N9 | A/duck/Germany/1/72(H2N9) |
| H3 | N8 | A/duck/Ukraine/63(H3N8) |
| H3 | N8 | A/duck/England/62(H3N8) |
| H3 | N2 | A/turkey/England/69(H3N2) |
| H4 | N6 | A/duck/Czechoslovakia/56(H4N6) |
| H4 | N3 | A/duck/Alberta/300/77(H4N3) |
| H5 | N3 | A/tern/South Africa/300/77(H4N3) |

| | | |
|---|---|---|
| H5 | N4 | A/Ethiopia/300/77(H6N6) |
| H5 | N9 | A/turkey/Ontario/7732/66(H5N9) |
| H5 | N1 | A/chick/Scotland/59(H5N1) |
| H6 | N2 | A/turkey/Massachusetts/3740/65(H6N2) |
| H6 | N8 | A/turkey/Canada/63(H6N8) |
| H6 | N5 | A/shearwater/Australia/72(H6N5) |
| H6 | N6 | A/jyotichinara/Ehiopia/73(H6N6) |
| H6 | N1 | A/duck/Germany/1868/68(H6N1) |
| H7 | N7 | A/fowl plague virus/Dutch/27(H7N7) |
| H7 | N1 | A/chick/Brescia/1902(H7N1) |
| H7 | N9 | A/chick/China/2013(H7N9) |
| H7 | N3 | A/turkey/England/639H7N3) |
| H7 | N1 | A/fowl plague virus/Rostock/34(H7N1) |
| H8 | N4 | A/turkey/Ontario/6118/68(H8N4) |
| H9 | N2 | A/turkey/Wisconsin/1/66(H9N2) |
| H9 | N6 | A/duck/Hong Kong/147/77(H9N6) |
| H9 | N6 | A/duck/Hong Kong/147/77(H9N6) |
| H9 | N8 | A/manishsurpur/Malawi/149/77(H9N8) |
| H9 | N7 | A/turkey/Scotland/70(H9N7) |
| H10 | N8 | A/quail/Italy/1117/65(H10N8) |
| H11 | N6 | A/duck/England/56(H11N6) |
| H11 | N9 | A/duck/Memphis/546/74(H11N9) |
| H12 | N5 | A/duck/Alberta/60/76/(H12N5) |
| H13 | N6 | A/gull/Maryland/704/77(H13N6) |
| H14 | N4 | A/duck/Gurjev/263/83(H14N4) |
| H15 | N9 | A/shearwater/Australia/2576/83(H15N9) |

## Spread

Avian influenza is most often spread by contact between infected and healthy birds, though can also be spread indirectly through contaminated equipment. The virus is found in secretions from the nostrils, mouth, and eyes of infected birds as well as their droppings. HPAI infection is spread to people often through direct contact with infected poultry, such as during slaughter or plucking. Though the virus can spread through airborne secretions, the disease itself is not an airborne disease. Highly pathogenic strains spread quickly among flocks and can destroy a flock within 28 hours; the less pathogenic strains may affect egg production but are much less deadly.

Although it is possible for humans to contract the avian influenza virus from birds, human-to-human contact is much more difficult without prolonged contact. However, public health officials are concerned that strains of avian flu may mutate to become easily transmissible between humans. Some strains of avian influenza are present in the intestinal tract of large numbers of shore birds and water birds, but these strains rarely cause human infection.

Five manmade ecosystems have contributed to modern avian influenza virus ecology: integrated indoor commercial poultry, range-raised commercial poultry, live poultry markets, backyard and hobby flocks, and bird collection and trading systems including cockfighting. Indoor commercial poultry has had the largest impact on the spread of HPAI, with the increase in HPAI outbreaks largely the result of increased commercial production since the 1990s.

## Village Poultry

In the early days of the HPAI H5N1 pandemic, village poultry and their owners were frequently implicated in disease transmission. Village poultry, also known as backyard and hobby flocks, are small flocks raised under extensive conditions and often allowed free range between multiple households. However, research has shown that these flocks pose less of a threat than intensively raised commercial poultry with homogenous genetic stock and poor biosecurity. Backyard and village poultry also do not travel great distances compared to transport of intensively raised poultry and contribute less to the spread of HPAI. This initial implication of Asian poultry farmers as one broad category presented challenges to prevention recommendations as commercial strategies did not necessarily apply to backyard poultry flocks.

## H5N1

The highly pathogenic influenza A virus subtype H5N1 is an emerging avian influenza virus that is causing global concern as a potential pandemic threat. It is often referred to simply as "bird flu" or "avian influenza", even though it is only one of many subtypes.

H5N1 has killed millions of poultry in a growing number of countries throughout Asia, Europe, and Africa. Health experts are concerned that the coexistence of human flu viruses and avian flu viruses (especially H5N1) will provide an opportunity for genetic material to be exchanged between species-specific viruses, possibly creating a new virulent influenza strain that is easily transmissible and lethal to humans. The mortality rate for humans with H5N1 is 60%.

Since first human H5N1 outbreak occurred in 1997, there has been an increasing number of HPAI H5N1 bird-to-human transmissions, leading to clinically severe and fatal human infections. Because a significant species barrier exists between birds and humans, the virus does not easily spread to humans, however some cases of infection are being researched to discern whether human-to-human transmission is occurring. More research is necessary to understand the pathogenesis and epidemiology of the H5N1 virus in humans. Exposure routes and other disease transmission characteristics, such as genetic and immunological factors that may increase the likelihood of infection, are not clearly understood.

The first known transmission of H5N1 to a human occurred in Hong Kong in 1997, when there was an outbreak of 18 human cases; 6 deaths were confirmed. None of the infected people worked with poultry. After culling all of the poultry in the area, no more cases were diagnosed. In 2006, the first human-to-human transmission likely occurred when 7 members of a family in Sumatra became infected after contact with a family member who had worked with infected poultry.

Although millions of birds have become infected with the virus since its discovery, 359 people have died from the H5N1 in twelve countries according to World Health Organization reports as of August 10, 2012.

As an example, the H5N1 outbreak in Thailand caused massive economic losses, especially among poultry workers. Infected birds were culled and slaughtered. The public lost confidence with the poultry products, thus decreasing the consumption of chicken products. This also elicited a ban from importing countries. There were, however, factors which aggravated the spread of the virus, including bird migration, cool temperature (increases virus survival) and several festivals at that time.

## Controversial Research

A study published in 2012 in Science Magazine reported on research findings that allowed for the airborne transmission of H5N1 in laboratory ferrets. The study identified the 5 mutations necessary for the virus to become airborne and immediately spiked controversy over the ethical implications of making such potentially dangerous information available to the general public. The study was allowed to remain available in its entirety, though remains a controversial topic within the scientific community.

The study in question, however, created airborne H5N1 via amino acid substitutions that largely mitigated the devastating effects of the disease. This fact was underscored by the 0% fatality rate among the ferrets infected via airborne transmission, as well as the fundamental biology underlying the substitutions. Flu viruses attach to host cells via the hemagluttinin proteins on their envelope. These hemagluttinin proteins bind to sialic acid receptors on host cells, which can fall into two categories. The sialic acid receptors can be either 2,3 or 2,6-linked, with the species of origin largely deciding receptor preference. In influenzas of avian origin 2,3-linkage is preferred, vs. influenzas of human origin in which 2,6-linkage is preferable. 2,3-linked SA receptors in humans are found predominantly in the lower respiratory tract, a fact that is the primary foundation for the deadliness of avian influenzas in humans, and also the key to their lack of airborne transmission. In the study that created an airborne avian influenza among ferrets it was necessary to switch the receptor preference of the host cells to those of 2,6-linkage, found predominantly in humans' upper respiratory tract, in order to create an infection that could shed aerosolized virus particles. Such an infection, however, must occur in the upper respiratory tract of humans, thus fundamentally undercutting the fatal trajectory of the disease.

## H7N9

Influenza A virus subtype H7N9 is a novel avian influenza virus first reported to have infected humans in 2013 in China. Most of the reported cases of human infection have resulted in severe respiratory illness. In the month following the report of the first case, more than 100 people had been infected, an unusually high rate for a new infection; a fifth of those patients had died, a fifth had recovered, and the rest remained critically ill. The World Health Organization (WHO) has identified H7N9 as "...an unusually dangerous virus for humans." As of June 30, 133 cases have been reported, resulting in the deaths of 43.

Research regarding background and transmission is ongoing. It has been established that many of the human cases of H7N9 appear to have a link to live bird markets. As of July 2013, there had been no evidence of sustained human-to-human transmission, however a study group headed by one of the world's leading experts on avian flu reported that several instances of human-to-human infection were suspected. It has been reported that H7N9 virus does not kill poultry, which will make surveillance much more difficult. Researchers have commented on the unusual prevalence of older males among H7N9-infected patients. While several environmental, behavioral, and biological explanations for this pattern have been proposed, as yet, the reason is unknown. Currently no vaccine exists, but the use of influenza antiviral drugs known as neuraminidase inhibitors in cases of early infection may be effective.

The number of cases detected after April fell abruptly. The decrease in the number of new human H7N9 cases may have resulted from containment measures taken by Chinese authorities, including closing live bird markets, or from a change in seasons, or possibly a combination of both factors. Studies indicate that avian influenza viruses have a seasonal pattern, thus it is thought that infections may pick up again when the weather turns cooler in China.

## In Domestic Animals

Several domestic species have been infected with and shown symptoms of H5N1 viral infection, including cats, dogs, ferrets, pigs, and birds.

## Birds

Attempts are made in the United States to minimize the presence of HPAI in poultry through routine surveillance of poultry flocks in commercial poultry operations. Detection of a HPAI virus may result in immediate culling of the flock. Less pathogenic viruses are controlled by vaccination, which is done primarily in turkey flocks (ATCvet codes: QI01AA23 (WHO) for the inactivated fowl vaccine, QI01CL01 (WHO) for the inactivated turkey combination vaccine).

## Seals

A recent strain of the virus is able to infect the lungs of seals.

## Cats

Avian influenza in cats can show a variety of symptoms and usually lead to death. Cats are able to get infected by either consuming an infected bird or by contracting the virus from another infected cat.

## Global Impact

In 2005, the formation of the International Partnership on Avian and Pandemic Influenza was announced in order to elevate the importance of avian flu, coordinate efforts, and improve disease reporting and surveillance in order to better respond to future pandemics. New networks of laboratories have emerged to detect and respond to avian flu, such as the Crisis Management Center for Animal Health, the Global Avian Influenza Network for Surveillance, OFFLU, and the Global Early Warning System for major animal diseases. After the 2003 outbreak, WHO member states have also recognized the need for more transparent and equitable sharing of vaccines and other benefits from these networks. Cooperative measures created in response to HPAI have served as a basis for programs related to other emerging and re-emerging infectious diseases.

HPAI control has also been used for political ends. In Indonesia, negotiations with global response networks were used to recentralize power and funding to the Ministry of Health. In Vietnam policymakers, with the support of the Food and Agriculture Organization of the United Nations (FAO), used HPAI control to accelerate the industrialization of livestock production for export by proposing to increase the portion of large-scale commercial farms and reducing the number of poultry keepers from 8 to 2 million by 2010.

## Stigma

Backyard poultry production was viewed as "traditional Asian" agricultural practices that contrasted with modern commercial poultry production and seen as a threat to biosecurity. Backyard production appeared to hold greater risk than commercial production due to lack of biosecurity and close contact with humans, though HPAI spread in intensively raised flocks was greater due to high density rearing and genetic homogeneity. Asian culture itself was blamed as the reason why certain interventions, such as those that only looked at placed-based interventions, would fail without looking for a multifaceted solutions.

## Indonesia

Press accounts of avian flu in Indonesia were seen by poultry farmers as conflating suspected cases while the public did see the accounts as informative, though many became de-sensitized to the idea of impending danger or only temporarily changed their poultry-related behavior. Rumors also circulated in Java in 2006. These tended to focus on bird flu being linked to big businesses in order to drive small farmers out of the market by exaggerating the danger of avian influenza, avian flu being introduced by foreigners to force Indonesians to purchase imported chicken and keep Indonesian chicken off the world market, and the government using avian flu as a ploy to attract funds from wealthy countries. Such rumors reflected concerns about big businesses, globalization, and a distrust of the national government in a country where "the amount of decentralization here is breathtaking" according to Steven Bjorge, a WHO epidemiologist in Jakarta in 2006.

In the context a decentralized national government that the public did not completely trust, Indonesian Health Minister Siti Fadilah Supari announced in December 2006 that her government would no longer be sharing samples of H5N1 collected from Indonesian patients. This decision came as a shock to the international community as it disrupted the Global Influenza Surveillance Network (GISN) coordinated by the WHO for managing seasonal and pandemic influenza. GISN is based on countries sharing virus specimens freely with the WHO which assesses and eventually sends these samples to pharmaceutical companies in order to produce vaccines that are sold back to these countries. Though this was initially seen as an attempt to protect national sovereignty at all costs, it was instead used for a domestic political struggle. Prior to Indonesia's dispute with the GISN, the Ministry of Health, already weak due to the decentralized nature the government, was experiencing further leakage of funding to state and non-state agencies due to global health interventions. By reasserting control over public health issues and funding by setting itself up as the sole Indonesian representative to the WHO, the Ministry of Health made itself a key player in the management of future international funds relating vaccine production and renegotiated benefits from global surveillance networks.

## Economic

Approximately 20% of the protein consumed in developing countries come from poultry. In the wake of the H5N1 pandemic, millions of poultry were killed. In Vietnam alone, over 50 million domestic birds were killed due to HPAI infection and control attempts. A 2005 report by the FAO totaled economic losses in South East Asia around US$10 billion. This had the greatest impact on small scale commercial and backyard producers relative to total assets compared to industrial chains which primarily experience temporary decreases in exports and loss of consumer confidence. Some governments did provide compensation for culled poultry, it was often far below market value (close to 30% of market value in Vietnam), while others such as Cambodia provide no compensation to farmers at all.

As poultry serves as a source of food security and liquid assets, the most vulnerable populations were poor small scale farmers. The loss of birds due to HPAI and culling in Vietnam led to an average loss of 2.3 months of production and US$69–108 for households where many have an income of $2 a day or less. The loss of food security for vulnerable households can be seen in the stunting of children under 5 in Egypt. Women are another population at risk as in most regions of the world, small flocks are tended to by women. Widespread culling also resulted in the decreased enrollment of girls in school in Turkey.

## Prevention

People who do not regularly come into contact with birds are not at high risk for contracting avian influenza. Those at high risk include poultry farm workers, animal control workers, wildlife biologists, and ornithologists who handle live birds. Organizations with high-risk workers should have an avian influenza response plan in place before any cases have been discovered. Biosecurity of poultry flocks is also important for prevention. Flocks should be isolated from outside birds, especially wild birds, and their waste; vehicles used around the flock should be regularly disinfected and not shared between farms; and birds from slaughter channels should not be returned to the farm.

With proper infection control and use of personal protective equipment (PPE), the chance for infection is low. Protecting the eyes, nose, mouth, and hands is important for prevention because these are the most common ways for the virus to enter the body. Appropriate personal protective equipment includes aprons or coveralls, gloves, boots or boot covers, and a head cover or hair cover. Disposable PPE is recommended. An N-95 respirator and unvented/indirectly vented safety goggles are also part of appropriate PPE. A powered air purifying respirator (PAPR) with hood or helmet and face shield is also an option.

Proper reporting of an isolated case can help to prevent spread. The Centers for Disease Control and Prevention (US) recommendation is that if a worker develops symptoms within 10 days of working with infected poultry or potentially contaminated materials, they should seek care and notify their employer, who should notify public health officials.

For future avian influenza threats, the WHO suggests a 3 phase, 5 part plan.

- Phase: Pre-pandemic
    - Reduce opportunities for human infection
    - Strengthen the early warning system
- Phase: Emergence of a pandemic virus
    - Contain or delay spread at the source

- Phase: Pandemic declared and spreading internationally
    - Reduce morbidity, mortality, and social disruption
    - Conduct research to guide response measures

Vaccines for poultry have been formulated against several of the avian H5N1 influenza varieties. Control measures for HPAI encourage mass vaccinations of poultry though The World Health Organization has compiled a list of known clinical trials of pandemic influenza prototype vaccines, including those against H5N1. In some countries still at high risk for HPAI spread, there is compulsory strategic vaccination though vaccine supply shortages remain a problem.

## For Village Poultry Farmers

Initial response to H5N1, a one size fits all recommendation was used for all poultry production systems, though measures for intensively raised birds were not necessarily appropriate for extensively raised birds. When looking at village poultry, it was first assumed that the household was the unit and that flocks did not make contact with other flocks, though more effective measures came into use when the epidemiological unit was the village.

Recommendations also involve restructuring commercial markets to improve biosecurity against avian influenza. Poultry production zoning is used to limit poultry farming to specific areas outside of urban environments while live poultry markets improve biosecurity by limiting the number of traders holding licenses and subjecting producers and traders to more stringent inspections. These recommendations in combination with requirements to fence and house all poultry, and limit free ranging flocks will eventually lead to fewer small commercial producers and backyard producers, costing livelihoods as they are unable to meet the conditions needed to participate.

A summary of reports to the World Organisation for Animal Health in 2005 and 2010 suggest that surveillance and under-reporting in developed and developing countries is still a challenge. Often, donor support can focus on HPAI control alone, while similar diseases such as Newcastle disease, acute fowl cholera, infectious laryngotracheitis, and infectious bursal disease still affect poultry populations. When HPAI tests come back negative, a lack of funded testing for differential diagnoses can leave farmers wondering what killed their birds.

Since traditional production systems require little investment and serve as a safety net for lower income households, prevention and treatment can been seen as less cost effective than letting a few birds die. Effective control not only requires prior agreements to be made with relevant government agencies, such as seen with Indonesia, they must also not unduly threaten food security.

## Culling

Culling is used in order to decrease the threat of avian influenza transmission by killing potentially infected birds. The FAO manual on HPAI control recommends a zoning strategy which begins with the identification of an infected area (IA) where sick or dead birds have tested positive. All poultry in this zone are culled while the area 1 to 5 km from the outer boundary of the IA is considered the restricted area (RA) placed under strict surveillance. 2 to 10 km from the RA is the control area (CA) that serves as a buffer zone in case of spread. Culling is not recommended beyond the IA unless there is evidence of spread. The manual, however, also provides examples of how control was carried out between 2004 and 2005 to contain H5N1 where all poultry was to be stamped out in a 3 km radius beyond the infected point and beyond that a 5 km radius where all fowl was to be vaccinated. This culling method was indiscriminate as a large proportion of the poultry inside these areas were small backyard flocks which did not travel great enough distances to carry infection to adjacent villages without human effort and may have not been infected at all. Between 2004 and 2005, over 100 million chickens were culled in Asia to contain H5N1.

The risk of mass culling of birds and the resulting economic impact led farmers who were reluctant to report sick poultry. The culls often preempted actual lab testing for H5N1 as avian flu policy justified sacrificing poultry as a safeguard against HPAI spread. In response to these policies, farmers in Vietnam between 2003 and 2004 became more and more unwilling to surrender apparently healthy birds to authorities and stole poultry destined for culls as it stripped poultry of their biosocial and economic worth. By the end of 2005, the government implemented a new policy that targeted high-risk flock in the immediate vicinity of infected farms and instituted voluntary culling with compensation in the case of a local outbreak.

Not only did culling result in severe economic impacts especially for small scale farmers, culling itself may be an ineffective preventative measure. In the short-term, mass culling achieves it goals of limiting the immediate spread of HPAI, it has been found to impede the evolution of host resistance which is important for the long-term success of HPAI control. Mass culling also selects for elevated influenza virulence and results in the greater mortality of birds overall. Effective culling strategies must be selective as well as considerate of economic impacts to optimize epidemiological control and minimize economic and agricultural destruction.

## People-poultry Relations

Prevention and control programs must take into account local understandings of people-poultry relations. In the past, programs that have focused on singular, place-based understandings of disease transmission have been ineffective. In the case of Northern Vietnam, health workers saw poultry as commodities with an environment that was under the control of people. Poultry existed in the context of farms, markets, slaughterhouses, an roads while humans were indirectly the primary transmitters of avian flu, placing the burden of disease control on people. However, farmers saw their free ranging poultry in an environment dominated by nonhuman forces that they could not exert control over. There were a host of non-human actors such as wild birds and weather patterns whose relationships with the poultry fostered the disease and absolved farmers of complete responsibility for disease control.

Attempts at singular, place-based controls sought to teach farmers to identify areas where their behavior could change without looking at poultry behaviors. Behavior recommendations by Vietnam's National Steering Committee for Avian Influenza Control and Prevention (NSCAI) were drawn from the FAO Principles of Biosecurity. These included restrictions from entering areas where poultry are kept by erecting barriers to segregate poultry from non-human contact, limits on human movement of poultry and poultry-related products ideally to transporters, and recommendations for farmers to wash hands and footwear before and after contact with poultry. Farmers, pointed to wind and environmental pollution as reasons poultry would get sick. NSCAI recommendations also would disrupt longstanding livestock production practices as gates impede sales by restricting assessment of birds by appearance and offend customers by limiting outside human contact. Instead of incorporating local knowledge into recommendations, cultural barriers were used as scapegoats for failed interventions. Prevention and control methods have been more effective when also considering the social, political, and ecological agents in play.

## Botulism

Botulism is a rare and potentially fatal illness caused by a toxin, produced by the bacterium *Clostridium botulinum*. The disease begins with weakness, trouble seeing, feeling tired, and trouble speaking. This may then be followed by weakness of the arms, chest muscles, and legs. The disease does not usually affect consciousness or cause a fever.

Botulism can be spread several different ways. The bacterial spores which cause it are common in both soil and water. They produce the botulinum toxin when exposed to low oxygen levels and certain temperatures. Foodborne botulism happens when food containing the toxin is eaten. Infant botulism happens when the bacteria develops in the intestines and releases the toxin. This typically only occurs in children less than six months old, as protective mechanisms develop after that time. Wound botulism is found most often among those who inject street drugs. In this situation, spores enter a wound, and in the absence of oxygen, release the toxin. It is not passed directly between people. The diagnosis is confirmed by finding the toxin or bacteria in the person in question.

Prevention is primarily by proper food preparation. The toxin, though not the organism, is destroyed by heating it to more than 85 °C (185 °F) for longer than 5 minutes. Honey can contain the organism, and for this reason, honey should not be fed to children under 12 months. Treatment is with an antitoxin. In those who lose their ability to breathe on their own, mechanical ventilation may be necessary for months. Antibiotics may be used for wound botulism. Death occurs in 5 to 10% of people. Botulism can affect many other animals. The word is from Latin, *botulus*, meaning sausage.

## Signs and Symptoms

The muscle weakness of botulism characteristically starts in the muscles supplied by the cranial nerves—a group of twelve nerves that control eye movements, the facial muscles and the muscles controlling chewing and swallowing. Double vision, drooping of both eyelids, loss of facial expression and swallowing problems may therefore occur. In addition to affecting the voluntary muscles, it can also cause disruptions in the autonomic nervous system. This is experienced as a dry mouth and throat (due to decreased production of saliva), postural hypotension (decreased blood pressure on standing, with resultant lightheadedness and risk of blackouts), and eventually constipation (due to decreased peristalsis). Some of the toxins (B and E) also precipitate nausea, vomiting, and difficulty with talking. The weakness then spreads to the arms (starting in the shoulders and proceeding to the forearms) and legs (again from the thighs down to the feet).

Severe botulism leads to reduced movement of the muscles of respiration, and hence problems with gas exchange. This may be experienced as dyspnea (difficulty breathing), but when severe can lead to respiratory failure, due to the buildup of unexhaled carbon dioxide and its resultant depressant effect on the brain. This may lead to coma and eventually death if untreated.

Clinicians frequently think of the symptoms of botulism in terms of a classic triad: bulbar palsy and descending paralysis, lack of fever, and clear senses and mental status ("clear sensorium").

## Infant Botulism

Infant botulism (also referred to as floppy baby syndrome) was first recognized in 1976, and is the most common form of botulism in the United States. There were 17 diagnosed cases of infant botulism in the United States in 2013. Infants are susceptible to infant botulism in the first year of life, with more than 90% of cases occurring in infants younger than six months. Infant botulism results from the ingestion of the *C. botulinum* spores, and subsequent colonization of the small intestine. The infant gut may be colonized when the composition of the intestinal microflora (normal flora) is insufficient to competitively inhibit the growth of *C. botulinum* and levels of bile acids (which normally inhibit clostridial growth) are lower than later in life.

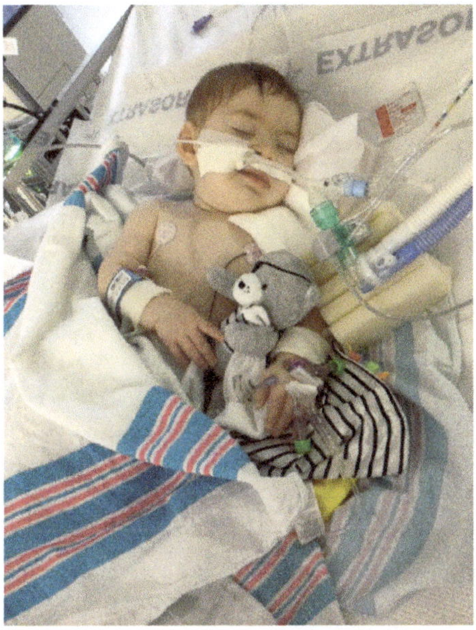


Infant with botulism. He is unable to move or open his eyes. His cry was weak. He is not asleep or sedated.

The growth of the spores releases botulinum toxin, which is then absorbed into the bloodstream and taken throughout the body, causing paralysis by blocking the release of acetylcholine at the neuromuscular junction. Typical symptoms of infant botulism include constipation, lethargy, weakness, difficulty feeding and an altered cry, often progressing to a complete descending flaccid paralysis. Although constipation is usually the first symptom of infant botulism, it is commonly overlooked.

Honey is a known dietary reservoir of *C. botulinum* spores and has been linked to infant botulism. For this reason honey is not recommended for infants less than one year of age. Most cases of infant botulism, however, are thought to be caused by acquiring the spores from the natural environment. *Clostridium botulinum* is a ubiquitous soil-dwelling bacterium. Many infant botulism patients have been demonstrated to live near a construction site or an area of soil disturbance.

Infant botulism has been reported in 49 of 50 US states, and cases have been recognized in 26 countries on five continents.

## Complications

Infant botulism has no long-term side effects, but can be complicated by hospital-acquired infections. The case fatality rate is less than 1% for hospitalized infants with botulism.

Botulism can result in death due to respiratory failure. However, in the past 50 years, the proportion of patients with botulism who die has fallen from about 50% to 7% due to improved supportive care. A patient with severe botulism may require mechanical ventilation (breathing support through a ventilator) as well as intensive medical and nursing care, sometimes for several months. The respiratory failure and paralysis that occur with severe botulism may require a person to be on a breathing machine (ventilator) for weeks or months. The person may require rehabilitation therapy after leaving the hospital.

## Cause

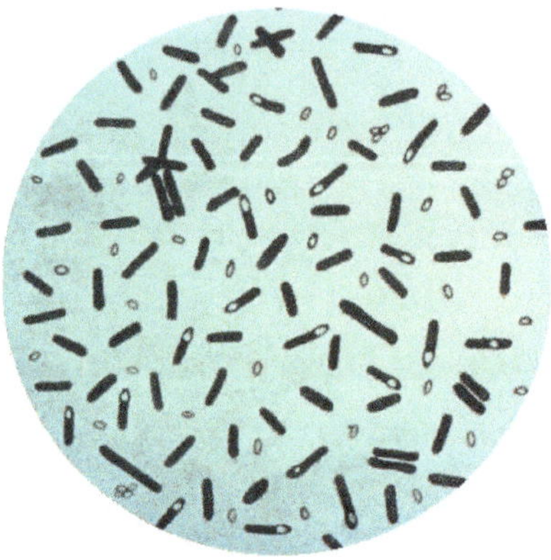

A photomicrograph of Clostridium botulinum bacteria.

*Clostridium botulinum* is an anaerobic, Gram positive, spore-forming rod. Botulinum toxin is one of the most powerful known toxins: about one microgram is lethal to humans when inhaled. It acts by blocking nerve function (neuromuscular blockade) through inhibition of the excitatory neurotransmitter acetylcholine's release from the presynaptic membrane of neuromuscular junctions in the somatic nervous system. This causes paralysis. Advanced botulism can cause respiratory failure by paralysing the muscles of the chest; this can progress to respiratory arrest.

In all cases, illness is caused by the botulinum toxin produced by the bacterium *C. botulinum* in anaerobic conditions and not by the bacterium itself. The pattern of damage occurs because the toxin affects nerves that fire (depolarise) at a higher frequency first.

Mechanisms of entry into the human body for botulinum toxin are described below.

## Colonization of the Gut

The most common form in Western countries is infant botulism. This occurs in infants who are colonized with the bacterium in the small intestine during the early stages of their lives. The bacterium then produces the toxin, which is absorbed into the bloodstream. The consumption of honey during the first year of life has been identified as a risk factor for infant botulism; it is a factor in a fifth of all cases. The adult form of infant botulism is termed *adult intestinal toxemia*, and is exceedingly rare.

## Food

Toxin that is produced by the bacterium within containers of food that have been improperly preserved is the most common cause of food-borne botulism. Fish that has been pickled without the salinity or acidity of brine that contains acetic acid and high sodium levels, as well as smoked fish stored at too high a temperature, presents a risk, as does improperly canned food.

Food borne botulism results from contaminated food in which *C. botulinum* spores have been allowed to germinate in low-oxygen conditions. This typically occurs in home-canned food substances and fermented uncooked dishes. Given that multiple people often consume food from the same source, it is common for more than a single person to be affected simultaneously. Symptoms usually appear 12–36 hours after eating, but can also appear within 2 hours to 10 days.

## Wound

Wound botulism results from the contamination of a wound with the bacteria, which then secrete the toxin into the bloodstream. This has become more common in intravenous drug users since the 1990s, especially people using black tar heroin and those injecting heroin into the skin rather than the veins. Wound botulism accounts for 29% of cases.

## Inhalation

Isolated cases of botulism have been described after inhalation by laboratory workers.

## Injection

Botulism has occurred after cosmetic use of inappropriate strengths of Botox.

## Mechanism

The toxin is the protein botulinum toxin produced under anaerobic conditions (where there is no oxygen) by the bacterium *Clostridium botulinum.*

*Clostridium botulinum* is a large anaerobic Gram-positive bacillus that forms subterminal endospores.

There are eight serological varieties of the bacterium denoted by the letters A to H. The toxin from all of these acts in the same way and produces similar symptoms: the motor nerve endings are prevented from releasing acetylcholine, causing flaccid paralysis and symptoms of blurred vision, ptosis, nausea, vomiting, diarrhea and/or constipation, cramps, and respiratory difficulty.

Botulinum toxin is broken into 8 neurotoxins (labeled as types A, B, C [C1, C2], D, E,

F, and G), which are antigenically and serologically distinct but structurally similar. Human botulism is caused mainly by types A, B, E, and (rarely) F. Types C and D cause toxicity only in other animals.

In October 2013, scientists released news of the discovery of type H, the first new botulism neurotoxin found in forty years. However, further information about type H has not been disclosed because of its potential for abuse as a lethal bioweapon and lack of a known antitoxin.

Some types produce a characteristic putrefactive smell and digest meat (types A and some of B and F); these are said to be proteolytic; type E and some types of B, C, D and F are nonproteolytic and can go undetected because there is no strong odor associated with them.

When the bacteria are under stress, they develop spores, which are inert. Their natural habitats are in the soil, in the silt that comprises the bottom sediment of streams, lakes and coastal waters and ocean, while some types are natural inhabitants of the intestinal tracts of mammals (e.g., horses, cattle, humans), and are present in their excreta. The spores can survive in their inert form for many years.

Toxin is produced by the bacteria when environmental conditions are favourable for the spores to replicate and grow, but the gene that encodes for the toxin protein is actually carried by a virus or phage that infects the bacteria. Unfortunately, little is known about the natural factors that control phage infection and replication within the bacteria.

The spores require warm temperatures, a protein source, an anaerobic environment, and moisture in order to become active and produce toxin. In the wild, decomposing vegetation and invertebrates combined with warm temperatures can provide ideal conditions for the botulism bacteria to activate and produce toxin that may affect feeding birds and other animals. Spores are not killed by boiling, but botulism is uncommon because special, rarely obtained conditions are necessary for botulinum toxin production from C. botulinum spores, including an anaerobic, low-salt, low-acid, low-sugar environment at ambient temperatures.

Botulinum inhibits the release within the nervous system of acetylcholine, a neurotransmitter, responsible for communication between motor neurons and muscle cells. All forms of botulism lead to paralysis that typically starts with the muscles of the face and then spreads towards the limbs. In severe forms, botulism leads to paralysis of the breathing muscles and causes respiratory failure. In light of this life-threatening complication, all suspected cases of botulism are treated as medical emergencies, and public health officials are usually involved to identify the source and take steps to prevent further cases from occurring.

Botulinum toxin A, C, and E cleave the SNAP-25, ultimately leading to paralysis.

## Diagnosis

For botulism in babies, diagnosis should be made on signs and symptoms. Confirmation of the diagnosis is made by testing of a stool or enema specimen with the mouse bioassay.

Physicians may consider diagnosing botulism if the patient's history and physical examination suggest botulism. However, these clues are often not enough to allow a diagnosis. Other diseases such as Guillain–Barré syndrome, stroke, and myasthenia gravis can appear similar to botulism, and special tests may be needed to exclude these other conditions. These tests may include a brain scan, cerebrospinal fluid examination, nerve conduction test (electromyography, or EMG), and an edrophonium chloride (Tensilon) test for myasthenia gravis. A definite diagnosis can be made if botulinum toxin is identified in the food, stomach or intestinal contents, vomit or feces. The toxin is occasionally found in the blood in peracute cases. Botulinum toxin can be detected by a variety of techniques, including enzyme-linked immunosorbent assays (ELISAs), electrochemiluminescent (ECL) tests and mouse inoculation or feeding trials. The toxins can be typed with neutralization tests in mice. In toxicoinfectious botulism, the organism can be cultured from tissues. On egg yolk medium, toxin-producing colonies usually display surface iridescence that extends beyond the colony.

## Prevention

Although the botulinum toxin is destroyed by thorough cooking over the course of a few minutes, the spore itself is not killed by the temperatures reached with normal sea-level-pressure boiling, leaving it free to grow and again produce the toxin when conditions are right.

A recommended prevention measure for infant botulism is to avoid giving honey to infants less than 12 months of age, as botulinum spores are often present. In older children and adults the normal intestinal bacteria suppress development of *C. botulinum*.

While commercially canned goods are required to undergo a "botulinum cook" in a pressure cooker at 121 °C (250 °F) for 3 minutes, and so rarely cause botulism, there have been notable exceptions such as the 1978 Alaskan salmon outbreak and the 2007 Castleberry's Food Company outbreak. Foodborne botulism is the rarest form though, accounting for only around 15% of cases (US) and has more frequently been from home-canned foods with low acid content, such as carrot juice, asparagus, green beans, beets, and corn. However, outbreaks of botulism have resulted from more unusual sources. In July 2002, fourteen Alaskans ate *muktuk* (whale meat) from a beached whale, and eight of them developed symptoms of botulism, two of them requiring mechanical ventilation.

Other, but much rarer sources of infection (about every decade in the US) include garlic or herbs stored covered in oil without acidification, chili peppers, improperly handled baked potatoes wrapped in aluminum foil, tomatoes, and home-canned or fermented fish.

When canning or preserving food at home, attention should be paid to hygiene, pressure, temperature, refrigeration and storage. When making home preserves, only acidic fruit such as apples, pears, stone fruits and berries should be bottled. Tropical fruit and tomatoes are low in acidity and must have some acidity added before they are bottled.

Oils infused with fresh garlic or herbs should be acidified and refrigerated. Potatoes which have been baked while wrapped in aluminum foil should be kept hot until served or refrigerated. Because the botulism toxin is destroyed by high temperatures, home-canned foods are best boiled for 10 minutes before eating. Metal cans containing food in which bacteria, possibly botulinum, are growing may bulge outwards due to gas production from bacterial growth; such cans should be discarded.

Any container of food which has been heat-treated and then assumed to be airtight which shows signs of not being so, e.g., metal cans with pinprick holes from rust or mechanical damage, should also be discarded. Contamination of a canned food solely with *C. botulinum* may not cause any visual defects (e.g. bulging). Only sufficient thermal processing during production should be used as a food safety control.

## Vaccine

There is a vaccine but its usefulness is unclear as it is associated with significant adverse effects. As of 2013 there are efforts ongoing to develop a better vaccine.

## Treatment

Botulism is generally treated with botulism antitoxin and supportive care.

Supportive care for botulism includes monitoring of respiratory function. Respiratory failure due to paralysis may require mechanical ventilation for 2 to 8 weeks, plus intensive medical and nursing care. After this time, paralysis generally improves as new neuromuscular connections are formed.

In some abnormal cases, physicians may try to remove contaminated food still in the digestive tract by inducing vomiting and/or using enemas. Wounds should be treated, usually surgically, to remove the source of the toxin-producing bacteria.

## Antitoxin

In adults, botulism can be treated by passive immunization with a horse-derived antitoxin, which blocks the action of the toxin circulating in the blood. A trivalent antitoxin containing antibodies raised against botulinum toxin types A, B, and E is used most commonly, however a heptavalent botulism antitoxin has also been developed and was approved by the U.S. FDA in 2013. In infants, horse-derived antitoxin is sometimes avoided for fear of infants developing serum sickness or lasting hypersensitivity to horse-derived proteins. To avoid this, a human-derived antitoxin has been developed and approved by the U.S. FDA in 2003 for the treatment of infant botulism. This human-derived antitoxin has been shown to be both safe and effective for the treatment of infant botulism. However, the danger of equine-derived antitoxin to infants has not been clearly established, and one study showed the equine-derived antitoxin to be both safe and effective for the treatment of infant botulism.

Trivalent (A,B,E) botulinum antitoxin is derived from equine sources utilizing whole antibodies (Fab and Fc portions). In the United States, this antitoxin is available from the local health department via the CDC. The second antitoxin, heptavalent (A,B,C,D,E,F,G) botulinum antitoxin, is derived from "despeciated" equine IgG antibodies which have had the Fc portion cleaved off leaving the F(ab')2 portions. This less immunogenic antitoxin is effective against all known strains of botulism where not contraindicated.

## Prognosis

The paralysis caused by botulism can persist for 2 to 8 weeks, during which supportive care and ventilation may be necessary to keep the patient alive. Botulism is fatal in 5 to 10% of people who are affected. However, if left untreated, botulism is fatal in 40 to 50% of cases.

The mortality rate for wound botulism is about 10%.

Infant botulism has no long-term side effects but can be complicated by nosocomial adverse events. The case fatality rate is less than 1% for hospitalized infants with botulism. In part because the vast majority of infants with botulism are hospitalized, the overall infant botulism mortality rate is about 1.3%.

## Epidemiology

Globally, botulism is fairly rare. In the United States, for example, an average of 145 cases are reported each year. Of these, roughly 65% are infant botulism, 20% are wound botulism, and 15% are foodborne. Infant botulism is predominantly sporadic and not associated with epidemics, but great geographic variability exists. From 1974 to 1996, for example, 47.2% of all infant botulism cases reported in the U.S. occurred in California.

Between 1990 and 2000, the Centers for Disease Control and Prevention reported 263 individual foodborne cases from 160 botulism events in the United States with a case-fatality rate of 4%. Thirty-nine percent (103 cases and 58 events) occurred in Alaska, all of which were attributable to traditional Alaska aboriginal foods. In the lower 49 states, home-canned food was implicated in 70 (91%) events with canned asparagus being the most numerous cause. Two restaurant-associated outbreaks affected 25 persons. The median number of cases per year was 23 (range 17–43), the median number of events per year was 14 (range 9–24). The highest incidence rates occurred in Alaska,

Idaho, Washington, and Oregon. All other states had an incidence rate of 1 case per ten million people or less.

The number of cases of food borne and infant botulism has changed little in recent years, but wound botulism has increased because of the use of black tar heroin, especially in California.

## United States

All data regarding botulism antitoxin releases and laboratory confirmation of cases in the US are recorded annually by the Centers for Disease Control and Prevention and published on their website.

- 1971 Bon Vivant botulism case On July 2, 1971, the U.S. Food and Drug Administration (FDA) released a public warning after learning that a New York man had died and his wife had become seriously ill due to botulism after eating a can of Bon Vivant vichyssoise soup.
- Between March 31 and April 6, 1977, 59 individuals developed type B botulism. All ill persons had eaten at the same Mexican restaurant in Pontiac, Michigan and all had consumed a hot sauce made with improperly home-canned jalapeño peppers, either by adding it to their food, or by eating a nacho that had had hot sauce used in its preparation. The full clinical spectrum (mild symptomatology with neurologic findings through life-threatening ventilatory paralysis) of type B botulism was documented.
- In April 1994, the largest outbreak of botulism in the United States since 1978 occurred in El Paso, Texas. Thirty persons were affected; 4 required mechanical ventilation. All ate food from a Greek restaurant. The attack rate among persons who ate a potato-based dip was 86% (19/22) compared with 6% (11/176) among persons who did not eat the dip (relative risk [RR] Å 13.8; 95% confidence interval [CI], 7.6–25.1). The attack rate among persons who ate an eggplant-based dip was 67% (6/9) compared with 13% (24/189) among persons who did not (RR Å 5.2; 95% CI, 2.9–9.5). Botulism toxin type A was detected from patients and in both dips. Toxin formation resulted from holding aluminum foil-wrapped baked potatoes at room temperature, apparently for several days, before they were used in the dips. Food handlers should be informed of the potential hazards caused by holding foil-wrapped potatoes at ambient temperatures after cooking.
- Beginning in late June 2007, 8 people contracted botulism poisoning by eating canned food products produced by Castleberry's Food Company in its Augusta, Georgia plant. It was later identified that the Castleberry's plant had serious production problems on a specific line of retorts that had under-processed the cans of food. These issues included broken cooking alarms, leaking water valves

and inaccurate temperature devices, all the result of poor management of the company. All of the victims were hospitalized and placed on mechanical ventilation. The Castleberry's Food Company outbreak was the first instance of botulism in commercial canned foods in the United States in over 30 years.

- One person died, 21 cases were confirmed, and 10 more were suspected in Lancaster, Ohio when a botulism outbreak occurred after a church potluck in April 2015. The suspected source was a salad made from home-canned potatoes.

## United Kingdom

The largest recorded outbreak of foodborne botulism in the United Kingdom occurred in June 1989. A total of 27 patients were affected; one patient died. Twenty-five of the patients had eaten one brand of hazelnut yogurt in the week before the onset of symptoms. This yogurt contained hazelnut conserve sweetened with aspartame rather than sugar. Control measures included the cessation of all yogurt production by the implicated producer, the withdrawal of the firm's yogurts from sale, the recall of cans of the hazelnut conserve, and advice to the general public to avoid the consumption of all hazelnut yogurts.

## Other Species

Botulism can occur in many vertebrates and invertebrates. Botulism has been reported in rats, mice, chicken, frogs, toads, goldfish, aplysia, squid, crayfish, drosophila, leeches, etc.

Death from botulism is common in waterfowl; an estimated 10,000 to 100,000 birds die of botulism annually. In some large outbreaks, a million or more birds may die. Ducks appear to be affected most often. Botulism also affects commercially raised poultry. In chickens, the mortality rate varies from a few birds to 40% of the flock.

Botulism seems to be relatively uncommon in domestic mammals; however, in some parts of the world, epidemics with up to 65% mortality are seen in cattle. The prognosis is poor in large animals that are recumbent.

In cattle, the symptoms may include drooling, restlessness, uncoordination, urine retention, dysphagia, and sternal recumbency. Laterally recumbent animals are usually very close to death. In sheep, the symptoms may include drooling, a serous nasal discharge, stiffness, and incoordination. Abdominal respiration may be observed and the tail may switch on the side. As the disease progresses, the limbs may become paralyzed and death may occur. Phosphorus-deficient cattle, especially in southern Africa, are inclined to ingest bones and carrion containing clostridial toxins and consequently suffer *lame sickness* or *lamsiekte*.

A recent study has demonstrated an effective vaccine against cattle botulism associated with Clostridium botulinum serotypes C and D.

The clinical signs in horses are similar to cattle. The muscle paralysis is progressive; it usually begins at the hindquarters and gradually moves to the front limbs, neck, and head. Death generally occurs 24 to 72 hours after initial symptoms and results from respiratory paralysis. Some foals are found dead without other clinical signs.

Domestic dogs may develop systemic toxemia after consuming C. botulinum type C exotoxin or spores within bird carcasses or other infected meat but are generally resistant to the more severe effects of Clostridium botulinum type C. Symptoms include flaccid muscle paralysis; dogs with breathing difficulties will require more intensive care monitoring. Muscle paralysis can lead to death due to cardiac and respiratory arrest.

Pigs are relatively resistant to botulism. Reported symptoms include anorexia, refusal to drink, vomiting, pupillary dilation, and muscle paralysis.

In poultry and wild birds, flaccid paralysis is usually seen in the legs, wings, neck and eyelids. Broiler chickens with the toxicoinfectious form may also have diarrhea with excess urates.

## Infectious Bursal Disease

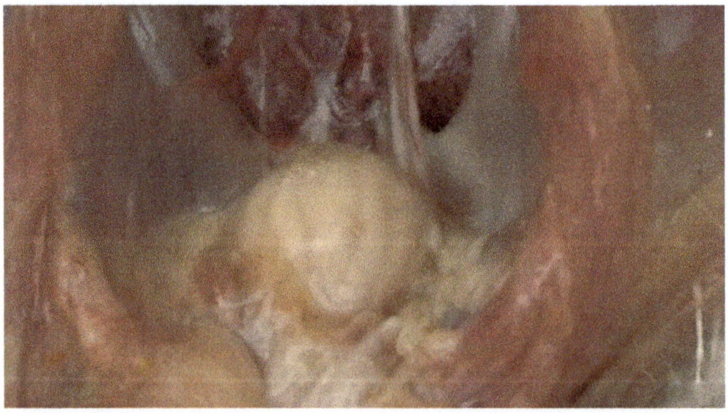

Enlarged bursa of Fabricius with yellowish peribursal oedema

Infectious bursal disease (also known as IBD, Gumboro Disease, Infectious Bursitis and Infectious Avian Nephrosis) is a highly contagious disease of young chickens caused by *infectious bursal disease virus* (IBDV), characterized by immunosuppression and mortality generally at 3 to 6 weeks of age. The disease was first discovered in Gumboro, Delaware in 1962. It is economically important to the poultry industry worldwide due to increased susceptibility to other diseases and negative interference with effective vaccination. In recent years, very virulent strains of IBDV (vvIBDV), causing severe mortality in chicken, have emerged in Europe, Latin America, South-East Asia, Africa and the Middle East. Infection is via the oro-fecal route, with affected bird excreting high levels of the virus for approximately 2 weeks after infection.

## Virology

IBDV is a double stranded RNA virus that has a bi-segmented genome and belongs to the genus *Avibirnavirus* of family *Birnaviridae*. There are two distinct serotypes of the virus, but only serotype 1 viruses cause disease in poultry. At least six antigenic subtypes of IBDV serotype 1 have been identified by *in vitro* cross-neutralization assay. Viruses belonging to one of these antigenic subtypes are commonly known as variants, which were reported to break through high levels of maternal antibodies in commercial flocks, causing up to 60 to 100 percent mortality rates in chickens. With the advent of highly sensitive molecular techniques, such as reverse transcription polymerase chain reaction (RT-PCR) and restriction fragment length polymorphism (RFLP), it became possible to detect the vvIBDV, to differentiate IBDV strains, and to use such information in studying the molecular epidemiology of the virus.

IBDV genome consists of two segments, A and B, which are enclosed within a nonenveloped icosahedral capsid. The genome segment B (2.9 kb) encodes VP1, the putative viral RNA polymerase. The larger segment A (3.2 kb) encodes viral proteins VP2, VP3, VP4, and VP5. Among them, VP2 protein contains important neutralizing antigenic sites and elicits protective immune response and most of the amino acid (AA) changes between antigenically different IBDVs are clustered in the hypervariable region of VP2. Thus, this hypervariable region of VP2 is the obvious target for the molecular techniques applied for IBDV detection and strain variation studies.

## Viral Structure

The IBDV capsid protein exhibits structural domains that show homology to those of the capsid proteins of some positive-sense single-stranded RNA viruses, such as the nodaviruses and tetraviruses, as well as the T=13 capsid shell protein of the *Reoviridae*. The T=13 shell of the IBDV capsid is formed by trimers of VP2, a protein generated by removal of the C-terminal domain from its precursor, pVP2. The trimming of pVP2 is performed on immature particles as part of the maturation process. The other major structural protein, VP3, is a multifunctional component lying under the T=13 shell that influences the inherent structural polymorphism of pVP2. The virus-encoded RNA-dependent RNA polymerase, VP1, is incorporated into the capsid through its association with VP3. VP3 also interacts extensively with the viral dsRNA genome.

## Pathogenesis

Clinical disease is associated to bird age with the greatest bursal mass, which occurs between 3 and 6 weeks of age. The greatest bursal mass is mostly a result of a large population of maturing IgM-bearing B-lymphocytes (lymphoblasts), the main target of infection. Young birds at around two to eight weeks of age that have highly active bursa of Fabricius are more susceptible to disease. Birds over eight weeks are resistant to challenge and will not show clinical signs unless infected by highly virulent strains.

Subclinical disease occurs in chickens infected before three weeks of age. At this age the B-lymphoblast population is smaller and the systemic effects are insufficient for generating clinical signs. However, the B-cell destruction is usually most severe in subclinically infected young, as virus will destroy a smaller population and most cells in one place (the bursa).

After ingestion, the virus destroys the lymphoid follicles in the bursa of Fabricius as well as the circulating B-cells in the secondary lymphoid tissues such as GALT (gut-associated lymphoid tissue), CALT (conjunctiva), BALT (Bronchial) caecal tonsils, Harderian gland, etc. Acute disease and death is due to the necrotizing effect of these viruses on the host tissues. Kidney failure is a common cause of mortality. If the bird survives and recovers from this phase of the disease, it remains immunocompromised which means it is more susceptible to other diseases.

## Clinical Signs

Disease may appear suddenly and morbidity typically reaches 100%. In the acute form birds are prostrated, debilitated and dehydrated. They produce a watery diarrhea and may have swollen feces-stained vent. Most of the flock is recumbent and have ruffled feathers. Mortality rates vary with virulence of the strain involved, the challenge dose, previous immunity, presence of concurrent disease, as well as the flock's ability to mount an effective immune response. Immunosuppression of very young chickens, less than three weeks of age, is possibly the most important outcome and may not be clinically detectable (subclinical). In addition, infection with less virulent strains may not show overt clinical signs, but birds that have bursal atrophy with fibrotic or cystic follicles and lymphocytopenia before six weeks of age, may be susceptible to opportunistic infection and may die of infection by agents that would not usually cause disease in immunocompetent birds.

## Diagnosis

A preliminary diagnosis can usually be made based on flock history, clinical signs and *post-mortem* (necropsy) examinations. However, definitive diagnosis can only be achieved by the specific detection and/or isolation and characterization of IBDV. Immunofluorescence or immunohistochemistry tests, based on anti-IBDV labelled antibodies, or in-situ hybridization, based on labelled complementary cDNA sequence probe, are useful for the specific detection of IBDV in infected tissues. RT-PCR (as mentioned above) was designed for the detection of IBDV genome, such as VP1 coding gene, with the possibility of PCR product sequences be determined for genetically comparing isolates and producing phylogenetic trees. Serological tests such as agar gel precipitation and ELISA, for detecting antibodies, are used for monitoring vaccine responses and might be additional information for diagnosis of infection of unvaccinated flocks.

Necropsy examination will usually show changes in the bursa of Fabricius such as swelling, oedema, haemorrhage, the presence of a jelly serosa transudate and eventually, bursal atrophy. Pathological changes, especially haemorrhages, may also be seen in the skeletal muscle, intestines, kidney and spleen.

## Treatment & Control

Peri-focal vaccination may not be effective for the combat of an outbreak, due to the rapidity of wild-IBDV spreading.

Passive immunity may protect against challenge with homologous IBDV, as does previous infection with homologous avirulent strains. Breeder flocks may be immunised against IBD so that they would transfer protective antibodies to their progenies, such as broiler and pullet chicks. Low-attenuated vaccine strains may cause damage to the bursa of Fabricius and immunosuppression in susceptible chicks. Biosecurity with adequate restriction to farm visitation and distancing from other flocks. Post outbreak hygiene measures may not be effective as the virus can survive for long periods in both housing and water.

Two enlarged bursae: yellowish grey (right) and haemorrhagic (left)

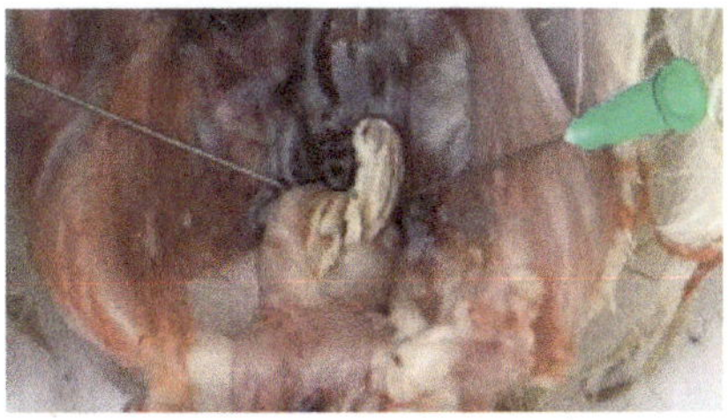

Caseous exudate in bursa of Fabricius

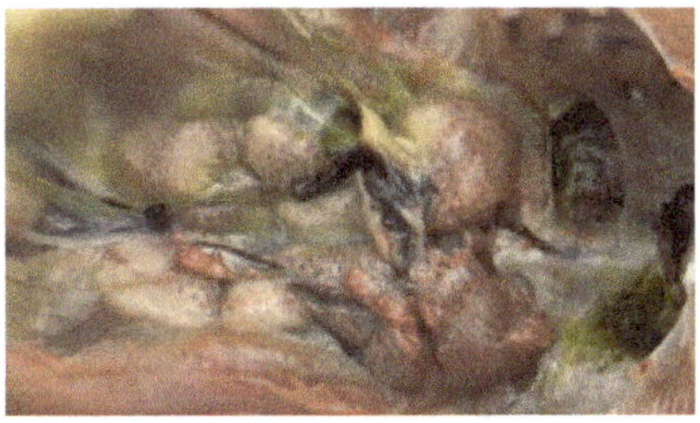

Lesions of kidneys

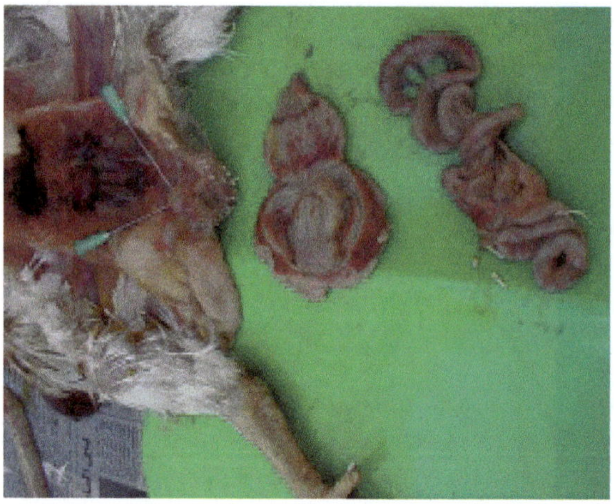

Haemorrhages in proventriculus and gizzard

## Toxoplasmosis

Toxoplasmosis is a parasitic disease caused by *Toxoplasma gondii*. Infections with toxoplasmosis usually cause no symptoms in adult humans. Occasionally there may be a few weeks or months of mild flu-like illness such as muscle aches and tender lymph nodes. In a small number of people, eye problems may develop. In those with a weak immune system, severe symptoms such as seizures and poor coordination may occur. If infected during pregnancy, a condition known as congenital toxoplasmosis may affect the child.

Toxoplasmosis is usually spread by eating poorly cooked food that contains cysts, exposure to infected cat feces, and from a mother to a child during pregnancy if the mother becomes infected. Rarely the disease may be spread by a blood transfusion. It is not otherwise spread between people. The parasite is only known to reproduce sexually in the cat family. However, it can infect most types of warm-blooded animals, including humans. Diagnosis is typically by testing the blood for antibodies or by testing the amniotic fluid for the parasite's DNA.

Prevention is by properly preparing and cooking food. It is also recommended that pregnant women not clean cat litter boxes. Treatment of otherwise healthy people is usually not needed. During pregnancy spiramycin or pyrimethamine/sulfadiazine and folinic acid may be used for treatment.

Up to half of the world's population are infected by toxoplasmosis but have no symptoms. In the United States about 23% are affected and in some areas of the world this is up to 95%. About 200,000 cases of congenital toxoplasmosis occur a year. Charles Nicolle and Louis Manceaux first described the organism in 1908. In 1941 transmission during pregnancy from a mother to a child was confirmed.

## Signs and Symptoms

Infection has three stages:

## Acute Toxoplasmosis

Acute toxoplasmosis is often asymptomatic in healthy adults. However, symptoms may manifest and are often influenza-like: swollen lymph nodes, headaches, fever, and fatigue, or muscle aches and pains that last for a month or more. Rarely will a human with a fully functioning immune system develop severe symptoms following infection. People with weakened immune systems are likely to experience headache, confusion, poor coordination, seizures, lung problems that may resemble tuberculosis or Pneumocystis jiroveci pneumonia (a common opportunistic infection that occurs in people with AIDS), or blurred vision caused by severe inflammation of the retina (ocular toxoplasmosis) Young children and immunocompromised people, such as those with HIV/AIDS, those taking certain types of chemotherapy, or those who have recently received an organ transplant, may develop severe toxoplasmosis. This can cause damage to the brain (encephalitis) or the eyes (necrotizing retinochoroiditis). Infants infected via placental transmission may be born with either of these problems, or with nasal malformations, although these complications are rare in newborns. The toxoplasmic trophozoites causing acute toxoplasmosis are referred to as tachyzoites, and are typically found in bodily fluids.

Swollen lymph nodes are commonly found in the neck or under the chin, followed by the armpits and the groin. Swelling may occur at different times after the initial infection, persist, and recur for various times independently of antiparasitic treatment. It is usually found at single sites in adults, but in children, multiple sites may be more common. Enlarged lymph nodes will resolve within one to two months in 60% of cases. However, a quarter of those affected take two to four months to return to normal, and 8% take four to six months. A substantial number (6%) do not return to normal until much later.

## Latent Toxoplasmosis

Due to its asymptomatic nature, it is easy for a host to become infected with *Toxoplasma gondii* and develop toxoplasmosis without knowing it. Although mild, flu-like symptoms occasionally occur during the first few weeks following exposure, infection with *T. gondii* produces no readily observable symptoms in healthy human adults. In most immunocompetent people, the infection enters a latent phase, during which only bradyzoites (tissue cysts) are present; these tissue cysts and even lesions can occur in the retinas, alveolar lining of the lungs (where an acute infection may mimic a *Pneumocystis jirovecii* infection), heart, skeletal muscle, and the central nervous system (CNS), including the brain. Cysts form in the CNS (brain tissue) upon infection with *T. gondii* and persist for the lifetime of the host. Most infants who are infected while in the womb have no symptoms at birth, but may develop symptoms later in life.

Reviews of serological studies have estimated that 30–50% of the global population has been exposed to and may be chronically infected with latent toxoplasmosis, although infection rates differ significantly from country to country. This latent state of infection has recently been associated with numerous disease burdens, neural alterations, and subtle gender-dependent behavioral changes in immunocompetent humans.

## Cutaneous Toxoplasmosis

While rare, skin lesions may occur in the acquired form of the disease, including roseola and erythema multiforme-like eruptions, prurigo-like nodules, urticaria, and maculopapular lesions. Newborns may have punctate macules, ecchymoses, or "blueberry muffin" lesions. Diagnosis of cutaneous toxoplasmosis is based on the tachyzoite form of *T. gondii* being found in the epidermis. It is found in all levels of the epidermis, is about 6 μm by 2 μm and bow-shaped, with the nucleus being one-third of its size. It can be identified by electron microscopy or by Giemsa staining tissue where the cytoplasm shows blue, the nucleus red.

## Cause

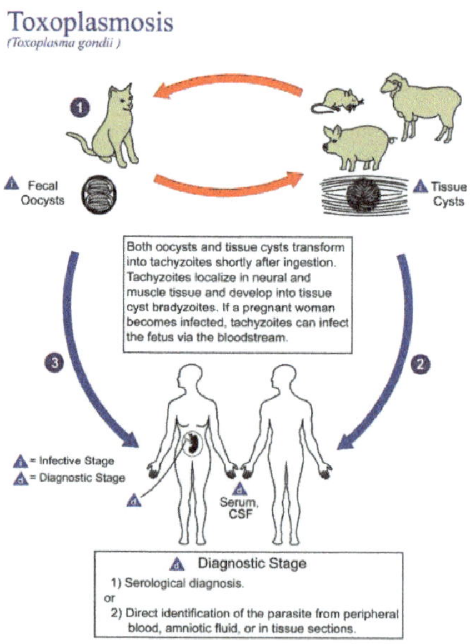


Lifecycle of *Toxoplasma gondii*

## Parasitology

In its life cycle, *T. gondii* adopts several forms. Tachyzoites are responsible for acute infection; they divide rapidly and spread through the tissues of the body. After proliferating, tachyzoites convert into bradyzoites, which take the form of latent intracellular tissue cysts that form mainly in the tissues of the muscles and brain. The transformation into cysts is in part triggered by the pressure of the host immune system. The bradyzoites are not responsive to antibiotics. The bradyzoites, once formed, can remain in the tissues for the lifespan of the host. In a healthy host, if some bradyzoites convert back into active tachyzoites, the immune system will quickly destroy them. However, in immunocompromised individuals, or in fetuses, which lack a developed immune system, the tachyzoites can run rampant and cause significant neurological damage.

The parasite's survival is dependent on a balance between host survival and parasite proliferation. *T. gondii* achieves this balance by manipulating the host's immune response, reducing the host's immune response and enhancing the parasite's reproductive advantage. Once it infects a normal host cell, it resists damage caused by the host's immune system, and changes the host's immune processes.

As it forces its way into the host cell, the parasite forms a parasitophorous vacuole (PV) membrane from the membrane of the host cell. The PV encapsulates the parasite, and is both resistant to the activity of the endolysosomal system, and can take control of the host's mitochondria and endoplasmic reticulum.

When first invading the cell, the parasite releases ROP proteins from the bulb of the rhoptry organelle. These proteins translocate to the nucleus and the surface of the PV membrane where they can activate STAT pathways to modulate the expression of cytokines at the transcriptional level, bind and inactivate PV membrane destroying IRG proteins, among other possible effects. Additionally, certain strains of T. *gondii* can secrete a protein known as GRA15, activating the NF-κB pathway, which upregulates the pro-inflammatory cytokine IL-12 in the early immune response, possibly leading to the parasite's latent phase. The parasite's ability to secrete these proteins depends on its genotype and affects its virulence.

The parasite also influences an anti-apoptotic mechanism, allowing the infected host cells to persist and replicate. One method of apoptosis resistance is by disrupting pro-apoptosis effector proteins, such as BAX and BAK. To disrupt these proteins, *T. gondii* causes conformational changes to the proteins, which prevent the proteins from being transported to various cellular compartments where they initiate apoptosis events. *T. gondii* does not, however, cause downregulation of the pro-apoptosis effector proteins.

*T. gondii* also has the ability to initiate autophagy of the host's cells. This leads to a decrease in healthy, uninfected cells, and consequently fewer host cells to attack the infected cells. Research by Wang *et al* finds that infected cells lead to higher levels of autophagosomes in normal and infected cells. Their research reveals that *T. gondii* causes host cell autophagy using a calcium-dependent pathway. Another study suggests that the parasite can directly affect calcium being released from calcium stores, which are important for the signalling processes of cells.

The mechanisms above allow *T. gondii* to persist in a host. Some limiting factors for the toxoplasma is that its influence on the host cells is stronger in a weak immune system and is quantity-dependent, so a large number of *T. gondii* per host cell cause a more severe effect. The effect on the host also depends on the strength of the host immune system. Immunocompetent individuals do not normally show severe symptoms or any at all, while fatality or severe complications can result in immunocompromised individuals.

It should be noted that since the parasite can change the host's immune response, it may also have an effect, positive or negative, on the immune response to other pathogenic threats. This includes, but is not limited to, the responses to infections by *Helicobacter felis*, *Leishmania major*, or other parasites, such as *Nippostrongylus brasiliensis*.

## Transmission

Toxoplasmosis is generally transmitted through the mouth when *toxoplama gondii* cysts are accidentally eaten. Congenital transmittance from the mother to fetus can also occur.

Oral transmission may occur through:

- Ingestion of raw or partly cooked meat, especially pork, lamb, or venison containing *Toxoplasma* cysts: Infection prevalence in countries where undercooked meat is traditionally eaten has been related to this transmission method. Tissue cysts may also be ingested during hand-to-mouth contact after handling undercooked meat, or from using knives, utensils, or cutting boards contaminated by raw meat.

- Ingestion of unwashed fruits or vegetables that have been in contact with contaminated soil containing infected cat feces.

- Ingestion of contaminated cat feces: This can occur through hand-to-mouth contact following gardening, cleaning a cat's litter box, contact with children's sandpits; the parasite can survive in the environment for over a year.

Cats excrete the pathogen in their feces for a number of weeks after contracting the disease, generally by eating an infected rodent. Even then, cat feces are not generally contagious for the first day or two after excretion, after which the cyst 'ripens' and becomes potentially pathogenic. In addition to cats, birds and mammals including human beings are also intermediate host of the spores and are involved in the transmission process. However the pathogenicity varies with the age and species involved in infection and the mode of transmission of T. gondii.

## Pregnancy Precautions

Congenital toxoplasmosis is a specific form of toxoplasmosis in which an unborn fetus is infected via the placenta. Congenital toxoplasmosis is associated with fetal death and abortion, and in infants, it is associated with neurologic deficits, neurocognitive deficits, and chorioretinitis. A positive antibody titer indicates previous exposure and immunity, and largely ensures the unborn fetus' safety. A simple blood draw at the first prenatal doctor visit can determine whether or not a woman has had previous exposure and therefore whether or not she is at risk. If a woman receives her first exposure to *T. gondii* while pregnant, the fetus is at particular risk.

Not much evidence exists around the effect of education before pregnancy to prevent congenital toxoplasmosis. However educating parents before the baby is born has been suggested to be effective because it may improve food, personal and pet hygiene. More research is needed to find whether antenatal education can reduce congenital toxoplasmosis.

For pregnant women with negative antibody titers, indicating no previous exposure to *T. gondii*, serology testing as frequent as monthly is advisable as treatment during pregnancy for those women exposed to *T. gondii* for the first time dramatically decreases the risk of passing the parasite to the fetus. Since a baby's immune system does not develop fully for the first year of life, and the resilient cysts that form throughout the body are very difficult to eradicate with antiprotozoans, an infection can be very serious in the young.

Despite these risks, pregnant women are not routinely screened for toxoplasmosis in most countries, for reasons of cost-effectiveness and the high number of false positives generated; Portugal, France, Austria, Uruguay, and Italy are notable exceptions, and some regional screening programmes operate in Germany, Switzerland and Belgium. As invasive prenatal testing incurs some risk to the fetus (18.5 pregnancy losses per toxoplasmosis case prevented), postnatal or neonatal screening is preferred. The exceptions are cases where fetal abnormalities are noted, and thus screening can be targeted.

Pregnant women should avoid handling raw meat, drinking raw milk (especially goat milk) and be advised to not eat raw or undercooked meat regardless of type. Because of the obvious relationship between *Toxoplasma* and cats it is also often advised to avoid exposure to cat feces, and refrain from gardening (cat feces are common in garden soil) or at least wear gloves when so engaged. Most cats are not actively shedding oocysts, since they get infected in the first six months of their life, when they shed oocysts for a short period of time (1–2 weeks.) However, these oocysts get buried in the soil, sporulate and remain infectious for periods ranging from several months to more than a year. Numerous studies have shown living in a household with a cat is not a significant risk factor for *T. gondii* infection, though living with several kittens has some significance.

In 2006, a Czech research team discovered women with high levels of toxoplasmosis antibodies were significantly more likely to have baby boys than baby girls. In most populations, the birth rate is around 51% boys, but women infected with *T. gondii* had up to a 72% chance of a boy. In mice, the sex ratio was higher in early latent toxoplasmosis and lower in later latent toxoplasmosis.

## Rodent Behavior

Infection with *T. gondii* has been shown to alter the behavior of mice and rats in ways thought to increase the rodents' chances of being preyed upon by cats. Infected rodents show a reduction in their innate aversion to cat odors; while uninfected mice and rats will generally avoid areas marked with cat urine or with cat body odor, this avoidance is reduced or eliminated in infected animals. Moreover, some evidence suggests this loss of aversion may be specific to feline odors: when given a choice between two predator odors (cat or mink), infected rodents show a significantly stronger preference to cat odors than do uninfected controls.

In rodents, *T. gondii*–induced behavioral changes occur through epigenetic remodeling in neurons associated with observed behaviors; for example, it modifies epigenetic methylation to induce hypomethylation of arginine vasopressin-related genes in the medial amygdala to greatly decrease predator aversion. Similar epigenetically-induced behavioral changes have also been observed in mouse models of addiction, where changes in the expression of histone-modifying enzymes via gene knockout or enzyme inhibition in specific neurons produced alterations in drug-related behaviors. Widespread histone-lysine acetylation in cortical astrocytes appears to be another epigenetic mechanism employed by *T. gondii*.

*T. gondii*-infected rodents show a number of behavioral changes beyond altered responses to cat odors. Rats infected with the parasite show increased levels of activity and decreased neophobic behavior. Similarly, infected mice show alterations in patterns of locomotion and exploratory behavior during experimental tests. These patterns include traveling greater distances, moving at higher speeds, accelerating for longer periods of time, and showing a decreased pause-time when placed in new arenas. Infected rodents have also been shown to have lower anxiety, using traditional models such as elevated plus mazes, open field arenas, and social interaction tests.

## Diagnosis

Diagnosis of toxoplasmosis in humans is made by biological, serological, histological, or molecular methods, or by some combination of the above. Toxoplasmosis can be difficult to distinguish from primary central nervous system lymphoma. It mimics several other infectious diseases so clinical signs are non-specific and are not sufficiently characteristic for a definite diagnosis. As a result, the diagnosis is made by a trial of therapy (pyrimethamine, sulfadiazine, and folinic acid (USAN: leucovorin)), if the drugs produce no effect clinically and no improvement on repeat imaging.

*T. gondii* may also be detected in blood, amniotic fluid, or cerebrospinal fluid by using polymerase chain reaction. *T. gondii* may exist in a host as an inactive cyst that would likely evade detection.

Serological testing can detect *T. gondii* antibodies in the blood serum, using methods including the Sabin–Feldman dye test (DT), the indirect hemagglutination assay, the indirect fluorescent antibody assay (IFA), the direct agglutination test, the latex agglutination test (LAT), the enzyme-linked immunosorbent assay (ELISA), and the immunosorbent agglutination assay test (IAAT).

The most commonly used tests to measure IgG antibody are the DT, the ELISA, the IFA, and the modified direct agglutination test. IgG antibodies usually appear within a week or two of infection, peak within one to two months, then decline at various rates. *Toxoplasma* IgG antibodies generally persist for life, and therefore may be present in the bloodstream as a result of either current or previous infection.

To some extent, acute toxoplasmosis infections can be differentiated from chronic infections using an IgG avidity test, which is a variation on the ELISA. In the first response to infection, toxoplasma-specific IgG has a low affinity for the toxoplasma antigen; in the following weeks and month, IgG affinity for the antigen increases. Based on the IgG avidity test, if the IgG in the infected individual has a high affinity, it means that the infection began three to five months before testing. This is particularly useful in congenital infection, where pregnancy status and gestational age at time of infection determines treatment.

In contrast to IgG, IgM antibodies can be used to detect acute infection, but generally not chronic infection. The IgM antibodies appear sooner after infection than the IgG antibodies and disappear faster than IgG antibodies after recovery. In most cases, *T. gondii*-specific IgM antibodies can first be detected approximately a week after acquiring primary infection, and decrease within one to six months; 25% of those infected are negative for *T. gondii*-specific IgM within seven months. However, IgM may be detectable months or years after infection, during the chronic phase, and false positives for acute infection are possible. The most commonly used tests for the measurement of IgM antibody are double-sandwich IgM-ELISA, the IFA test, and the immunosorbent agglutination assay (IgM-ISAGA). Commercial test kits often have low specificity, and the reported results are frequently misinterpreted.

## Congenital Toxoplasmosis

Recommendations for the diagnosis of congenital toxoplasmosis include: prenatal diagnosis based on testing of amniotic fluid and ultrasound examinations; neonatal diagnosis based on molecular testing of placenta and cord blood and comparative mother-child serologic tests and a clinical examination at birth; and early childhood diagnosis based on neurologic and ophthalmologic examinations and a serologic survey during the first year of life. During pregnancy, serological testing is recommended at three week intervals.

Even though diagnosis of toxoplasmosis heavily relies on serological detection of specific anti-*Toxoplasma* immunoglobulin, serological testing has limitations. For example, it may fail to detect the active phase of *T. gondii* infection because the specific anti-*Toxoplasma* IgG or IgM may not be produced until after several weeks of infection. As a result, a pregnant woman might test negative during the active phase of *T. gondii* infection leading to undetected and therefore untreated congenital toxoplasmosis. Also, the test may not detect *T. gondii* infections in immunocompromised patients because the titers of specific anti-*Toxoplasma* IgG or IgM may not rise in this type of patient.

Many PCR-based techniques have been developed to diagnose toxoplasmosis using clinical specimens that include amniotic fluid, blood, cerebrospinal fluid, and tissue biopsy. The most sensitive PCR-based technique is nested PCR, followed by hybridization of PCR products. The major downside to these techniques is that they are time consuming and do not provide quantitative data.

Real-time PCR is useful in pathogen detection, gene expression and regulation, and allelic discrimination. This PCR technique utilizes the 5' nuclease activity of *Taq* DNA polymerase to cleave a nonextendible, fluorescence-labeled hybridization probe during the extension phase of PCR. A second fluorescent dye, e.g., 6-carboxy-tetramethyl-rhodamine, quenches the fluorescence of the intact probe. The nuclease cleavage of the hybridization probe during the PCR releases the effect of quenching resulting in an increase of fluorescence proportional to the amount of PCR product, which can be monitored by a sequence detector.

Toxoplasmosis cannot be detected with immunostaining. Lymph nodes affected by *Toxoplasma* have characteristic changes, including poorly demarcated reactive germinal centers, clusters of monocytoid B cells, and scattered epithelioid histiocytes.

## Treatment

Treatment is often only recommended for people with serious health problems, such as people with HIV whose CD4 counts are under 200 cells/mm$^3$, because the disease is most serious when one's immune system is weak. Trimethoprim/sulfamethoxazole is the drug of choice to prevent toxoplasmosis, but not for treating active disease. A new study (May 2012) shows a promising new way to treat the active and latent form of this disease using two endochin-like quinolones.

### Acute

The medications prescribed for acute toxoplasmosis are the following:

- Pyrimethamine — an antimalarial medication
- Sulfadiazine — an antibiotic used in combination with pyrimethamine to treat toxoplasmosis
    - Combination therapy is usually given with folic acid supplements to reduce incidence of thrombocytopaenia.

  - Combination therapy is most useful in the setting of HIV.
- Clindamycin
- Spiramycin – an antibiotic used most often for pregnant women to prevent the infection of their children

(other antibiotics, such as minocycline, have seen some use as a salvage therapy).

### Latent

In people with latent toxoplasmosis, the cysts are immune to these treatments, as the antibiotics do not reach the bradyzoites in sufficient concentration.

The medications prescribed for latent toxoplasmosis are:

- Atovaquone – an antibiotic that has been used to kill *Toxoplasma* cysts inside AIDS patients
- Clindamycin – an antibiotic that, in combination with atovaquone, seemed to optimally kill cysts in mice

### Congenital Toxoplasmosis

When a pregnant woman is diagnosed with acute toxoplasmosis, amniocentesis can be used to determine whether the fetus has been infected or not. When a pregnant woman develops acute toxoplasmosis, the tachyzoites have approximately a 30% chance of entering the placental tissue, and from there entering and infecting the fetus. As gestational age at the time of infection increases, the chance of fetal infection also increases.

If the parasite has not yet reached the fetus, spiramycin can help to prevent placental transmission. If the fetus has been infected, the pregnant woman can be treated with pyrimethamine and sulfadiazine, with folinic acid, after the first trimester. They are given after the first trimester and with folinic acid because pyrimethamine has an antifolate effect, and lack of folic acid can interfere with fetal brain formation and cause thrombocytopaenia. Infection in earlier gestational stages correlates with poorer fetal and neonatal outcomes, particularly when the infection is untreated.

## Epidemiology

*T. gondii* infections occur throughout the world, although infection rates differ significantly by country. For women of childbearing age, a survey of 99 studies within 44 countries found the areas of highest prevalence are within Latin America (about 50–80%), parts of Eastern and Central Europe (about 20–60%), the Middle East (about 30-50%), parts of Southeast Asia (about 20–60%), and parts of Africa (about 20–55%).

In the United States, data from the National Health and Nutrition Examination Survey (NHANES) from 1999 to 2004 found 9.0% of US-born persons 12–49 years of age were seropositive for IgG antibodies against *T. gondii*, down from 14.1% as measured in the NHANES 1988–1994. In the 1999–2004 survey, 7.7% of US-born and 28.1% of foreign-born women 15–44 years of age were *T. gondii* seropositive. A trend of decreasing seroprevalence has been observed by numerous studies in the United States and many European countries.

The protist responsible for Toxoplasmosis is *T. gondii*. There are three major types of T. gondii responsible for the patterns of Toxoplasmosis throughout the world. There are types I, II, and III. These three types of *T. gondii* have differing effects on certain hosts, mainly mice and humans due to their variation in genotypes.

- Type I: virulent in mice and humans, seen in AIDS patients.
- Type II: non-virulent in mice, virulent in humans (mostly Europe and North America), seen in AIDS patients.
- Type III: non-virulent in mice, virulent mainly in animals but seen to a lesser degree in humans as well.

Current serotyping techniques can only separate type I or III from type II parasites.

Because the parasite poses a particular threat to fetuses when it is contracted during pregnancy, much of the global epidemiological data regarding *T. gondii* comes from seropositivity tests in women of childbearing age. Seropositivity tests look for the presence of antibodies against *T. gondii* in blood, so while seropositivity guarantees one has been exposed to the parasite, it does not necessarily guarantee one is chronically infected.

## History

*Toxoplasma gondii* was first described in 1908 by Nicolle and Manceaux in Tunisia, and independently by Splendore in Brazil. Splendore reported the protozoan in a rabbit, while Nicolle and Manceaux identified it in a North African rodent, the gundi (Ctenodactylus gundi). In 1909 Nicolle and Manceaux differentiated the protozoan from *Leishmania*. Nicolle and Manceaux then named it *Toxoplasma gondii* after the curved shape of its infectious stage (Greek root 'toxon'= bow).

The first recorded case of congenital toxoplasmosis was in 1923, but it was not identified as caused by *T. gondii*. Janků (1923) described in detail the autopsy results of an 11-month-old boy who had presented to hospital with hydrocephalus. The boy had classic marks of toxoplasmosis including chorioretinitis (inflammation of the choroid and retina of the eye). Histology revealed a number of "sporocytes", though Janků did not identify these as *T. gondii*.

It was not until 1937 that the first detailed scientific analysis of *T. gondii* took place using techniques previously developed for analyzing viruses. In 1937 Sabin and Olitsky analyzed *T. gondii* in laboratory monkeys and mice. Sabin and Olitsky showed that *T. gondii* was an obligate intracellular parasite and that mice fed *T. gondii*-contaminated tissue also contracted the infection. Thus Sabin and Olitsky demonstrated *T. gondii* as a pathogen transmissible between animals.

*T. gondii* was first identified as a human pathogen in 1939. Wolf, Cowen and Paige identified *T. gondii* infection in an infant girl delivered full-term by Caesarean section. The infant developed seizures and had chorioretinitis in both eyes at three days. The infant then developed encephalomyelitis and died at one month of age. Wolf, Cowen and Paige isolated *T. gondii* from brain tissue lesions. Intracranial injection of brain and spinal cord samples into mice, rabbits and rats produced encephalitis in the animals. Wolf, Cowen and Page reviewed additional cases and concluded that *T. gondii* produced recognizable symptoms and could be transmitted from mother to child.

The first adult case of toxoplasmosis was reported in 1940 with no neurological signs. Pinkerton and Weinman reported the presence of *Toxoplasma* in a 22-year-old man from Peru who died from a subsequent bacterial infection and fever.

In 1948, a serological dye test was created by Sabin and Feldman based on the ability of the patient's antibodies to alter staining of *Toxoplasma*. The Sabin Feldman Dye Test is now the gold standard for identifying *Toxoplasma* infection.

Transmission of *Toxoplasma* by eating raw or undercooked meat was demonstrated by Desmonts et al. in 1965 Paris. Desmonts observed that the therapeutic consumption of raw beef or horse meat in a tuberculosis hospital was associated with a 50% per year increase in *Toxoplasma* antibodies. This means that more *T. gondii* was being transmitted through the raw meat.

In 1974, Desmonts and Couvreur showed that infection during the first two trimesters produces most harm to the fetus, that transmission depended on when mothers were infected during pregnancy, that mothers with antibodies before pregnancy did not transmit the infection to the fetus, and that spiramycin lowered the transmission to the fetus.

*Toxoplasma* gained more attention in the 1970s with the rise of immune-suppressant treatment given after organ or bone marrow transplants and the AIDS epidemic of the 1980s. Patients with lowered immune system function are much more susceptible to disease.

## Society and Culture

### "Crazy Cat-lady Syndrome"

"Crazy cat-lady syndrome" is a term coined by news organizations to describe scientific findings that link the parasite *Toxoplasma gondii* to several mental disorders and behavioral problems. Although researches found that cat ownership does not strongly increase the risk of a *T. gondii* infection in pregnant women, the suspected correlation between cat ownership in childhood and later development of schizophrenia suggest that further studies are needed to determine a risk factor for children. The term *crazy cat-lady syndrome* draws on both stereotype and popular cultural reference. It was originated as instances of the aforementioned afflictions were noted amongst the populace. Cat lady is a cultural stereotype of a woman, often a spinster, who compulsively hoards and dotes upon cats. Jaroslav Flegr (biologist) is a proponent of the theory that toxoplasmosis affects human behaviour.

## Notable Cases

- Arthur Ashe (tennis player) developed neurological problems from toxoplasmosis (and was later found to be HIV-positive).
- Merritt Butrick (actor) was HIV-positive and died from toxoplasmosis as a result of his already-weakened immune system.
- Pedro Zamora (reality television personality and HIV/AIDS activist) was diagnosed with toxoplasmosis as a result of his immune system being weakened by HIV.
- Prince François, Count of Clermont (pretender to the throne of France); his disability caused him to be overlooked in the line of succession.
- Leslie Ash (actress) contracted toxoplasmosis in the second month of pregnancy.
- Sebastian Coe (British middle-distance runner).
- Martina Navratilova suffered from toxoplasmosis during the 1982 US Open.
- Louis Wain (artist) was famous for painting cats; he later developed schizophrenia, which some believe was due to toxoplasmosis resulting from his prolonged exposure to cats.
- The Wealdstone Raider suffered toxoplasmosis for much of his childhood and adolescence.

## Popular Culture

### Literature

- In Irvine Welsh's debut novel *Trainspotting* (1993), the character Matty dies of toxoplasmosis. In the film adaptation of the novel, it is the HIV-positive character Tommy who dies of toxoplasmosis.

## Other Animals

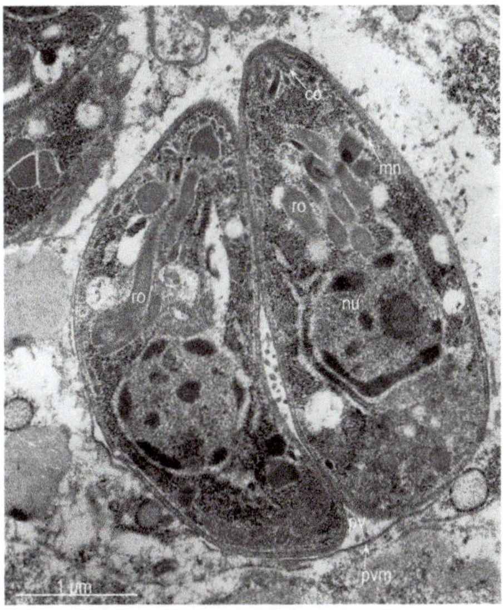


*Toxoplasma gondii* infects virtually all warm-blooded animals; these tachyzoites were found in a bird

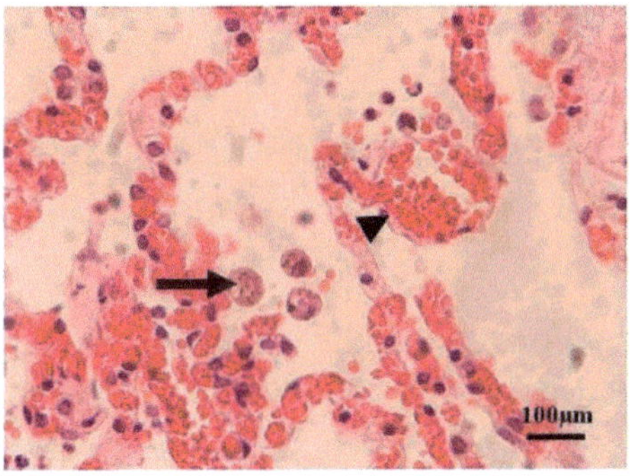


*Toxoplasma gondii* in the lung of a Giant panda. Arrow: macrophages containing tachyzoites

Although *T. gondii* has the capability of infecting virtually all warm-blooded animals, susceptibility and rates of infection vary widely between different genera and species. Rates of infection in populations of the same species can also vary widely due to differences in location, diet, and other factors.

Although infection with *T. gondii* has been noted in several species of Asian primates, seroprevalence of *T. gondii* antibodies was found for the first time in toque macaques (*Macaca sinica*) that are endemic to the island of Sri Lanka.

## Livestock

Among livestock, pigs, sheep and goats have the highest rates of chronic *T. gondii* infection. The prevalence of *T. gondii* in meat-producing animals varies widely both within and among countries, and rates of infection have been shown to be dramatically influenced by varying farming and management practices. For instance, animals kept outdoors or in free-ranging environments are more at risk of infection than animals raised indoors or in commercial confinement operations.

In the United States, the percentage of pigs harboring viable parasites has been measured (via bioassay in mice or cats) to be as high as 92.7% and as low as 0%, depending on the farm or herd. Surveys of seroprevalence (*T. gondii* antibodies in blood) are more common, and such measurements are indicative of the high relative seroprevalence in pigs across the world. Along with pigs, sheep and goats are among the most commonly infected livestock of epidemiological significance for human infection. Prevalence of viable *T. gondii* in sheep tissue has been measured (via bioassay) to be as high as 78% in the United States, and a 2011 survey of goats intended for consumption in the United States found a seroprevalence of 53.4%.

Due to a lack of exposure to the outdoors, chickens raised in large-scale indoor confinement operations are not commonly infected with *T. gondii*. Free-ranging or backyard-raised chickens are much more commonly infected. A survey of free-ranging chickens in the United States found its prevalence to be 17%–100%, depending on the farm. Because chicken meat is generally cooked thoroughly before consumption, poultry is not generally considered to be a significant risk factor for human *T. gondii* infection.

Although cattle and buffalo can be infected with *T. gondii*, the parasite is generally eliminated or reduced to undetectable levels within a few weeks following exposure. Tissue cysts are rarely present in buffalo meat or beef, and meat from these animals is considered to be low-risk for harboring viable parasites.

Horses are considered resistant to chronic *T. gondii* infection. However, viable cells have been isolated from US horses slaughtered for export, and severe human toxoplasmosis in France has been epidemiologically linked to the consumption of horse meat.

## Domestic Cats

In 1942, first case of feline toxoplasmosis was diagnosed and reported in a domestic cat in Middletown, NY. The investigators were able to isolate oocysts from feline feces and it was also noted that the oocysts can be infectious for up to 12 months in the environment.

The seroprevalence of *T. gondii* in domestic cats, worldwide, has been estimated to be around 30–40%. In the United States, no official national estimate has been made, but local surveys have shown levels varied between 16% and 80%. A 2012 survey of 445 purebred pet cats and 45 shelter cats in Finland found an overall seroprevalence of 48.4%. A 2010 survey of feral cats from Giza, Egypt, found an overall seroprevalence of 97.4%. Another survey from Colombia showed the seroprevalence of 89.3% whereas a Chinese study showed seroprevalence of 2.1%.

*T. gondii* infection rates in domestic cats vary widely depending on the cats' diets and lifestyles. Feral cats that hunt for their food are more likely to be infected than domestic cats. The prevalence of *T. gondii* in cat populations depends on the availability of infected birds and small mammals, but often this prey is abundant.

Most infected cats will shed oocysts only once in their lifetimes, for a period of about one to two weeks. Although this period of shedding is quite transient, millions of oocysts can be shed, with each oocyst capable of spreading and surviving for months. An estimated 1% of cats at any given time are actively shedding oocysts.

It is difficult to control the cat population with the infected oocysts due to lack of an effective vaccine. This remains a challenge in most cases and the programs that are readily available are questionable in efficacy.

## Marine Mammals

A University of California, Davis study of dead sea otters collected from 1998 to 2004 found toxoplasmosis was the cause of death for 13% of the animals. Proximity to freshwater outflows into the ocean was a major risk factor. Ingestion of oocysts from cat feces is considered to be the most likely ultimate source. Surface runoff containing wild cat feces and litter from domestic cats flushed down toilets are possible sources of oocysts. The parasites have been found in dolphins and whales. Researchers Black and Massie believe anchovies, which travel from estuaries into the open ocean, may be helping to spread the disease.

## Giant Panda

*Toxoplasma gondii* has been reported as the cause of death of a Giant panda kept in a zoo in China, who died in 2014 of acute gastroenteritis and respiratory disease. Although seemingly anecdotal, this report emphasizes that all warm-bloodied species are likely to be infected by *T. gondii*, including endangered species such as the Giant panda.

## Research

Chronic infection with *T. gondii* has traditionally been considered asymptomatic in people with normal immune function.

Some evidence suggests latent infection may subtly influence a range of human behaviors and tendencies, and infection may alter the susceptibility to or intensity of a number of affective, psychiatric, or neurological disorders. Research has linked toxoplasmosis with schizophrenia.

Latent *T. gondii* infection in humans has been associated with a higher incidence of automobile accidents, potentially due to impaired psychomotor performance or enhanced risk-taking personality profiles. Moreover, correlations have been found between positive antibody titers to *T. gondii* and OCD, Parkinson's disease, Alzheimer's disease, suicide in people with mood disorders, and bipolar disorder. Positive antibody titers to *T. gondii* have been shown to be not correlative with major depression or dysthymia. Although there is a correlation between T. gondii infection and many psychological disorders, scientists are still trying to find the cause on a cellular level.

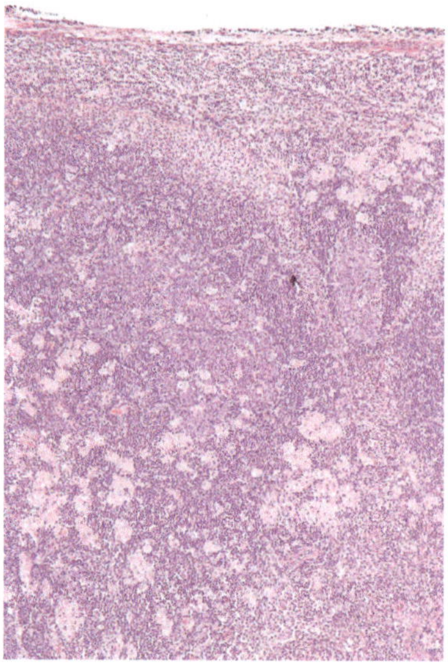

Micrograph of a lymph node showing the characteristic changes of toxoplasmosis (scattered epithelioid histiocytes (pale cells), monocytoid cells (top-center of image), large germinalcenters (left of image)) H&E stain

A 2016 study using the Dunedin cohort found that "there was little evidence that T. gondii was related to increased risk of psychiatric disorder, poor impulse control, personality aberrations or neurocognitive impairment".

## Schizophrenia

Some evidence links *T. gondii* to schizophrenia. Two 2012 meta-analysis found the rates of antibodies to *T.gondii* in schizophrenia was 2.7 times higher than that in control populations. *T.gondii* antibody positivity was therefore considered an intermediate risk factor in relation to other known risk factors.

Cautions that were noted include that the antibody tests do not detect toxoplasmosis directly, most people with schizophrenia do not have antibodies for toxoplasmosis, and publication bias might exist.

The majority of these studies tested people already diagnosed with schizophrenia for *T. gondii* antibodies, associations between *T. gondii* and schizophrenia have been found prior to the onset of schizophrenia symptoms. Studies attempting to explain the sex differences in schizophrenia onset have suggested it can be accounted for by a second peak of *T.gondii* infection incidence occurring from ages 25–30 observed in females only. People with schizophrenia and *T. gondii* antibodies show a higher mortality rate than schizophrenics testing seronegative. Although a mechanism supporting the association between schizophrenia and *T.gondii* infection is unclear, studies have attempted to discern a molecular mechanism to explain the correlation.

Studies have demonstrated that antipsychotic drugs used in the treatment of schizophrenia appear to inhibit the replication of T. gondii tachyzoites in cell culture.

### Traffic Accidents

In addition to a correlation with psychological disorders such as OCD, schizophrenia, and bipolar disorder, *T. gondii* can also lead to infected people having a higher risk of being in a car accident than uninfected individuals. A study in the Czech Republic found that latent toxoplasmosis patients were involved in accidents 2.65 times more often than people without toxoplasmosis infection. The risk also increased with significantly higher amounts of *T. gondii* antibodies in the host cells.

### Limitations of Correlation Analysis

In most of the current studies where positive correlations have been found between *T. gondii* antibody titers and certain behavioral traits or neurological disorders, *T. gondii* seropositivity tests are conducted after the onset of the examined disease or behavioral trait; that is, it is often unclear whether infection with the parasite increases the chances of having a certain trait or disorder, or if having a certain trait or disorder increases the chances of becoming infected with the parasite. Groups of individuals with certain behavioral traits or neurological disorders may share certain behavioral tendencies that increase the likelihood of exposure to and infection with *T. gondii*; as a result, it is difficult to confirm causal relationships between *T. gondii* infections and associated neurological disorders or behavioral traits. Provided there is in fact an etiological link between *T. gondii* and schizophrenia, studies have yet to determine why some individuals with latent toxoplasmosis develop schizophrenia while others do not, however, some plausible explanations include differing genetic susceptibility, parasite strain differences, and differences in the route of the acquired *T.gondii* infection.

### Multiple Sclerosis

There is a negative association between an infection with the parasite T. gondii and multiple sclerosis, therefore, researchers have concluded that toxoplasmosis infection could be considered a protective factor.

## Avian Reovirus

Avian reoviruses belong to the genus *Orthoreovirus*, and *Reoviridae* family. They are non-enveloped viruses that undergo replication in the cytoplasm of infected cells. It has icosahedral symmetry and contains a double-shelled arrangement of surface protein. Virus particles can range between 70–80 nm. Morphologically, the virus is a double stranded RNA virus that is composed of ten segments. The genome and proteins that are encoded by the genome can be separated into three different sizes ranging from small, medium, or large. Of the eleven proteins that are encoded for by the genome, two are nonstructural, while the remaining nine are structural.

Avian reoviruses can withstand a pH range of 3.0–9.0. Ambient temperatures are suitable for the survival of these viruses, which become inactive at 56 °C in less than an hour. Common areas where this virus can survive include galvanized metal, glass, rubber, feathers, and wood shavings. Avian reovirus can survive for up to ten days on these common areas in addition to up to ten weeks in water.

Cultivation and observation of the effects of avian reovirus is most often performed in chicken embryos. If infected into the yolk sac, the embryo will succumb to death accompanied by hemorrhaging of the embryos and cause the foci on the liver to appear yellowish-green. There are several primary chicken cell cultures/areas that are susceptible to avian reoviruses, which include the lungs, liver, kidney, and fibroblasts of the chick embryo. Of the following susceptible areas, liver cells from the chick embryo have been found to be the most sensitive for primary isolation from clinical material. Typically, the CPE effect of avian reoviruses is the production of syncytia. CPE, or cytopathic effects are the visible changes in a host cell that takes place because of viral infection. Syncytia is a single cell or cytoplasmic mass containing several nuclei, formed by fusion of cells or by division of nuclei.

## Genome Structure

The complete genome sequence of avian reovirus is 23,494 bp long and is broken up into 10 segments. The full lengths of each segment are L1 (3,959 nt), L2 (3,830 nt), L3 (3,907 nt), M1 (2,283 nt), M2 (2,158 nt), M3 (1,996 nt), S1 (1,643 nt), S2 (1,324 nt), S3 (1,202 nt), and S4 (1,192 nt). There are 10 proteins that are also involved in the composition of this virus. The lengths of these structural and non-structural proteins are λa (1,293 amino acids), λB (1,259 aa), λC (1,285 aa), μA (732 aa), μB (676 aa), μNS (635 aa), σC (326 aa), σA (416 aa), σB (367 aa), and σNS (367 aa).

## Virus-host Interactions

Alpha and beta interferons belong to a family of multifunctional cytokines, which are expressed and activated/distributed by fibroblasts and leukocytes in response to infections. In an experiment where primary cultures of chicken embryo fibroblasts were treated with alpha/beta interferon, results indicated that this interaction induced an antiviral state that strongly inhibited vaccinia and vesicular stomatitis virus (VSV) replication, but didn't have an effect on avian reovirus, specifically the replication of the S1133 strain. Instead, extracts of cells that were infected with avian-reovirus were able to block the activation of dsRNA-dependent enzymes, enabling the virus to relieve the translation-inhibitory activity of dsRNA. Also, protein σA, which is a core protein of the S1133 strain, binds to dsRNA irreversibly. Removing this protein from this particular viral strain disables the posttranslational capacity of infected cells. Research has shown that this core protein may possibly disrupt the IFN-induced cellular response against avian reovirus through the blockage of enzyme pathways that are dependent on dsRNA.

In addition to the possible disruption of IFN-induced cellular responses, chicken embryo fibroblasts that are infected with strain S1133 have displayed significant cytopathic effects. Strain S1133 led to the activation of the intracellular death in the early stages of infection. Although viral gene expression is not necessary, intracellular viral uncoating is the significant process that induces apoptosis. The ability of avian reovirus to induce apoptosis is not restricted to a distinct cell type of virus strain due to the fact that several different isolates of this pathogen could induce apoptosis in different mammalian and avian cell lines. Apoptosis can also be induced in cells that are infected with UV-irradiated reovirus virions and cells infected with ribavirin-treated avian reovirus, which suggests that viral mRNA synthesis and crucial steps that occur during viral replication are not necessary for induction of apoptosis in infected cells. Disregarding infectivity, the number of inoculated virus particles is the significant factor that leads to apoptosis from avian reovirus. Induced apoptosis is stopped when intracellular viral uncoating is blocked.

## Epidemiology

Avian reovirus can be transmitted both vertically and horizontally. Egg transmission has been observed after experimental infection, however the rate of transmission is very low. Congenitally infected chicks are assumed to act as a nucleus of infection for the rest of the hatch due to their high probability of becoming infected through the fecal-oral route or through the respiratory tract. Infection may also enter through the exposure of broken skin of the feet or legs of the chickens, where it then can become established in the hock joints. Once the infection has entered the body, avian reovirus can survive in the tissues of chickens for many weeks. Resistance to reovirus infections in chickens is directly related to age. Chicks that are infected one day after birth are more prone to experimentally synthesized tenosynovitis/arthritis than those that were infected at two weeks or older. Chicks that were infected a day after birth also displayed a development of more severe joint lesions and higher intestinal virus titers than those who were infected at two weeks of age. Infectious agents, which increase the pathogenicity of reovirus in the joints of chicken, include *Mycoplasma synoviae*, *Staphylococcus aureus*, infectious bursal disease virus, and chicken anemia virus. Avian reoviruses are also typically resistant to certain disinfectants.

## Virus Distribution

The most common route of infection occurs orally, and occasionally through the respiratory tract from the nucleus of congenitally infected hatch mates. Experimental infection of adult chickens through the esophagus, nasal passages, or trachea caused the virus to spread throughout all areas of the enteric, respiratory tract, reproductive tract, and the hock and tendon joints. In an experiment, avian reovirus was recovered from mononuclear, plasma, and erythrocyte cell fractions of blood within 30 hours of infection in young chicks. After 3–5 days, the infection spread throughout the whole body. The main site of viral replication was observed in the enteric tract. A different study of early pathogenesis in chicks infected with virus one day after birth displayed that the bursa of Fabricius and the epithelial cells of the small intestines are the main sites of infection and portal of entry of the virus which quickly disseminates to other organs within a 24- to 48-hour window post infection. The tibiotarsal-tarsometatarsal (hock joint) was the site where virus replication caused the most severe damage and in some extreme cases, tendon rupture.

## Immune Responses

### Humoral Antibodies

The sera of birds infected with avian reovirus display circulating antibodies through the validation of ELISA, agar gel immunodiffusion, indirect immunofluorescence (IIF), and virus neutralization (VN). Virus neutralization identifies type-specific antibodies, which allows for differentiation between strains of viruses through their distinct antigens. The other four tests detect group antigens.

### Maternal Antibodies

Maternal antibodies have displayed protection against the development of microscopic lesions of tenosynovitis in chicks that have been infected one day after birth. Protection provided by maternal antibodies has served as a foundation of breeder vaccination.

### Cell-mediated Immunity

An experiment that used monoclonal antibodies that were specific for chick Ia (a chicken class II major histocompatibility complex antigen), and T and B-lymphocytes were observed to determine its effects on cellular infiltrates during the development of reovirus arthritis. Within the synovium collected, plasma cells and T-lymphocytes were the primary inflammatory cells present. During the acute-phase, CD8 cells were present in low numbers. The most activity was observed during the subacute phase with an increase in CD8 and CD4 lymphocytes. At this phase, clusters of IgM-positive, B-cells, T-cells, and plasma cells were also observed. During the chronic phase, a high amount of CD4 T- cells and a few IgM-positive B-cells were observed.

## Immunosuppression

Infections of chickens with avian reovirus doesn't disrupt the functional capabilities of their T-cells, but induces suppressor macrophages that inhibit T-cell function.

## Diseases

There are several diseases that are caused by avian reovirus, which includes, avian arthritis/tenosynovitis, runting-stunting syndrome, and blue wing disease in chickens. Blue wing disease affects young broiler chickens and has an average mortality rate of 10%. It causes intramuscular and subcutaneous hemorrhages and atrophy of the spleen, bursa of Fabricius, and thymus. When young chickens are experimentally infected with avian reovirus, it is spread rapidly throughout all tissues. This virus is spread most frequently in the skin and muscles, which is also the most obvious site for lesions. Avian arthritis causes significant lameness in joints, specifically the hock joints. In the most severe cases, viral arthritis has caused the tendon to rupture. Chickens that have contracted runting-stunting syndrome cause a number of individuals in a flock to appear noticeably small due to its delayed growth. Diseased chicks are typically pale, dirty, wet, and may have a distending abdomen. Some individuals may display "helicopter-like" feathers in their wings and other feather abnormalities. The virus has also been shown to cause osteoporosis.

## Diagnostic Methods

Although infection of avian reovirus is spread worldwide, it is rarely the sole cause of a disease. For chickens, the most common manifestation of the disease is joint/limb lameness. Confirming infection of avian reovirus can be detected through an ELISA test by using and observing the expression of σC and σB proteins. However, isolating and identifying reoviruses from tissue samples is very time consuming. Isolation is most successfully attained through inoculation of material into chick embryo cultures or fertile chicken eggs. Inoculation of embryonic eggs through the yolk sac has shown that the virus usually kills the embryos within 5 or 6 days post inoculation. Analyzing the samples, the embryos appeared hemorrhagic and necrotic lesions on the liver were present. (Jones, Onunkwo, 1978). There have also been approaches to identify avian reoviruses molecularly by observing infected tissues with dot-blot hybridization, PCR, and a combination of PCR and RFLP. This combination allows for the reovirus strain to be typed.

## Prevention and Control Methods

Vaccines are available (ATCvet codes: QI01AA04 (WHO) for the inactivated vaccine, QI01AD10 (WHO) for the live vaccine, plus various combinations). Given that avian reovirus infections are widespread, the viruses are relatively resistant outside the host, and that vertical and horizontal transmission occurs, eradicating avian reovirus infection in commercial chicken flocks is very unlikely. In addition, absence of detectable seroconversion and failure to detect virus in cloacal swabs are unreliable indicators of resisting infection, or transmission via the egg. Thus, the most proactive and successful approach to controlling this disease is through vaccination. Since chicks are more prone to being detrimentally affected by the disease right after hatching, vaccine protocols that use live and killed vaccines are designed to provide protection during the very early stages of life. This approach has been accomplished through active immunity after early vaccination and a live vaccine or passive immunity from maternal antibodies followed with vaccination of the breeder hens. Currently, efforts toward administering inactivated or live vaccines to breeding stock to allow passive immunity to the offspring via the yolk are being taken.

## Erysipelas

Erysipelas is an acute infection typically with a skin rash, usually on any of the legs and toes, face, arms, and fingers. It is an infection of the upper dermis and superficial lymphatics, usually caused by beta-hemolytic group A *Streptococcus* bacteria on scratches or otherwise infected areas. Erysipelas is more superficial than cellulitis, and is typically more raised and demarcated.

### Signs and Symptoms

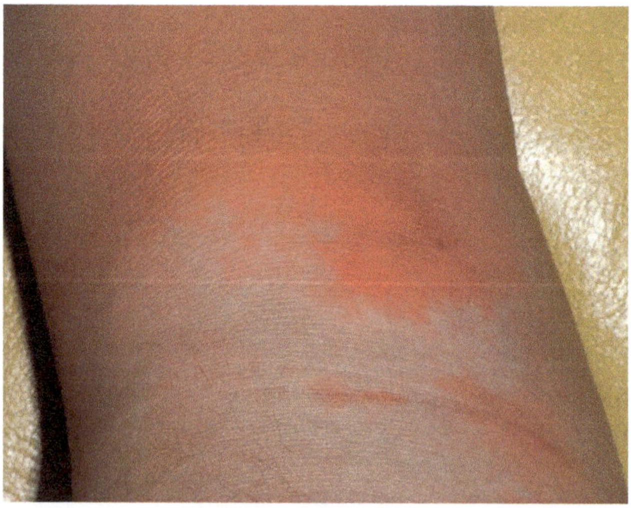

Erysipelas on an arm

Affected individuals typically develop symptoms including high fevers, shaking, chills, fatigue, headaches, vomiting, and general illness within 48 hours of the initial infection. The erythematous skin lesion enlarges rapidly and has a sharply demarcated, raised edge. It appears as a red, swollen, warm, and painful rash, similar in consistency to an orange peel. More severe infections can result in vesicles (pox or insect bite-like marks), blisters, and petechiae (small purple or red spots), with possible skin necrosis (death). Lymph nodes may be swollen, and lymphedema may occur. Occasionally, a red streak extending to the lymph node can be seen.

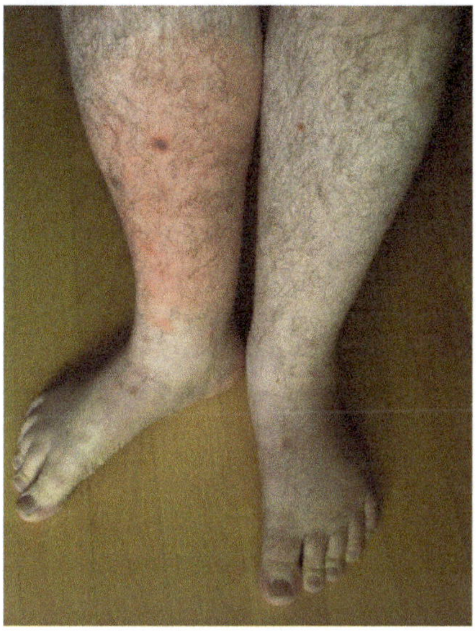

Erysipelas on a leg

The infection may occur on any part of the skin, including the face, arms, fingers, legs, and toes; it tends to favour the extremities. Fat tissue and facial areas, typically around the eyes, ears, and cheeks, are most susceptible to infection. Repeated infection of the extremities can lead to chronic swelling (lymphangitis).

## Cause

Most cases of erysipelas are due to *Streptococcus pyogenes* (also known as beta-hemolytic group A streptococci), although non-group A streptococci can also be the causative agent. Beta-hemolytic, non-group A streptococci include *Streptococcus agalactiae*, also known as group B strep or GBS. Historically, the face was most affected; today, the legs are affected most often. The rash is due to an exotoxin, not the *Streptococcus* bacteria, and is found in areas where no symptoms are present; e.g., the infection may be in the nasopharynx, but the rash is found usually on the face and arms.

Erysipelas infections can enter the skin through minor trauma, insect bites, dog bites, eczema, athlete's foot, surgical incisions and ulcers and often originate from streptococci bacteria in the subject's own nasal passages. Infection sets in after a small scratch or abrasion spreads, resulting in toxaemia.

Erysipelas does not affect subcutaneous tissue. It does not release pus, only serum or serous fluid. Subcutaneous edema may lead the physician to misdiagnose it as cellulitis, but the style of the rash is much more well circumscribed and sharply marginated than the rash of cellulitis.

## Risk Factors

This disease is most common among the elderly, infants, and children. People with immune deficiency, diabetes, alcoholism, skin ulceration, fungal infections, and impaired lymphatic drainage (e.g., after mastectomy, pelvic surgery, bypass grafting) are also at increased risk.

## Diagnosis

This disease is diagnosed mainly by the appearance of well-demarcated rash and inflammation. Blood cultures are unreliable for diagnosis of the disease, but may be used to test for sepsis. Erysipelas must be differentiated from herpes zoster, angioedema, contact dermatitis, and diffuse inflammatory carcinoma of the breast.

Erysipelas can be distinguished from cellulitis by its raised advancing edges and sharp borders. Elevation of the antistreptolysin O titer occurs after around 10 days of illness.

## Treatment

Depending on the severity, treatment involves either oral or intravenous antibiotics, using penicillins, clindamycin, or erythromycin. While illness symptoms resolve in a day or two, the skin may take weeks to return to normal.

Because of the risk of reinfection, prophylactic antibiotics are sometimes used after resolution of the initial condition. However, this approach does not always stop reinfection.

## Prognosis

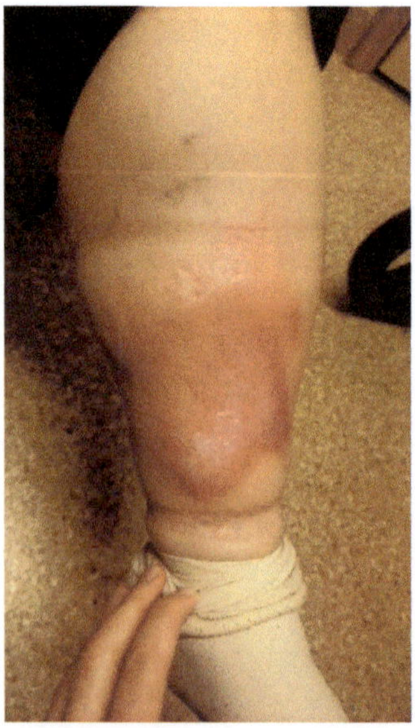

Recurrent erysipelas on an edematous leg

The disease prognosis includes:

- Spread of infection to other areas of body can occur through the bloodstream (bacteremia), including septic arthritis. Glomerulonephritis can follow an episode of streptococcal erysipelas or other skin infection, but not rheumatic fever.
- Recurrence of infection: Erysipelas can recur in 18–30% of cases even after antibiotic treatment. A chronic state of recurrent erysipelas infections can occur with several predisposing factors including alcoholism, diabetes, and tinea pedis (athlete's foot). Another predisposing factor is chronic cutaneous edema, such as can in turn be caused by venous insufficiency or heart failure.
- Lymphatic damage
- Necrotizing fasciitis, commonly known as "flesh-eating" bacterial infection, is a potentially deadly exacerbation of the infection if it spreads to deeper tissue.

## Notable Cases

*Apocryphal/Fictional*

- Richard Wagner, German composer, also suffered from erysipelas, although he died of a heart attack.
- In D. H. Lawrence's famous novel *Sons and Lovers* one of the major characters in the novel, William Morel (who is the main character Paul's older brother) dies quickly from the complications of erysipelas in conjunction with pneumonia.
- In Anton Chekhov's 1892 short story *Ward No. 6*, erysipelas is among the afflictions suffered by the patients committed to a poorly run mental illness facility in a small town in tsarist Russia.

*Fatal, in order of death*

- Archibald Douglas, 6th Earl of Angus, (d. 1557), Scottish nobleman active in the reigns of James V and Mary, Queen of Scots
- John of the Cross, Spanish saint and priest (d. 1591)
- Michiel de Ruyter, Dutch admiral in the Anglo-Dutch wars, contracted from injuries sustained from a cannonball. (d. 1676)
- Christina, Queen of Sweden (d. 1689)
- Anne, Queen of Great Britain and Ireland (d. 1714)
- Norborne Berkeley, baron de Botetourt, Royal Governor of Virginia (d. 1770)

- Princess Amelia of the United Kingdom, daughter of George III of the United Kingdom (1783–1810)
- Grand Duchess Catherine Pavlovna of Russia, daughter of Tsar Paul I of Russia and wife of King William I of Württemberg (d. 1819)
- William Wirt, United States Attorney General and U.S. presidential candidate (d. 1834)
- Charles Lamb, English writer and essayist (d. 1834)
- Barbara Hofland, English children's writer and novelist (d. 1844)
- Pope Gregory XVI (d. 1846)
- Mary Lyon, American women's education pioneer (d. 1849)
- John Herbert White, youngest son of James S. and Ellen G. White, co-founders of the Seventh-day Adventist church (d. 1860)
- Frederick VII of Denmark, king of Denmark (d. 1863)
- John Stuart Mill, English political philosopher (d. 1873)
- Marcus Clarke, Australian journalist, poet, playwright and novelist, who wrote "For the Term of His Natural Life", died age 35 (d. 1881)
- John Brown, Scottish personal servant and companion to Queen Victoria (d. 1883)
- Pat Killen, American heavyweight boxer, died at age 29 while in hiding in Chicago from police after assaulting two men (d. 1891)
- Samuel Augustus Ward, American organist, composer, teacher, businessman (d. 1903)
- James Anthony Bailey, American circus ringmaster (d. 1906)
- George Herbert, 5th Earl of Carnarvon (d. 1923), English aristocrat known as the financial backer of the search for and excavation of Tutankhamun's tomb in the Valley of the Kings. His death led to the story of the Curse of Tutankhamun.
- Miller Huggins, American baseball player and manager (d. 1929)
- Father Solanus Casey, American Capuchin priest declared "venerable" by the Roman Catholic Church (d. 1957)

*Recovered*

- Lenin suffered an infection in London, and party leadership was exercised by Martov until he recovered.

### Other Animals

Erysipelas is also the name given to an infection in animals caused by the bacterium *Erysipelothrix rhusiopathiae*. *E. rhusiopathiae* can also infect humans, but in that case the infection is known as erysipeloid.

## Haemoproteus

*Haemoproteus* is a genus of protozoa that are parasitic in birds, reptiles and amphibians. Its name is derived from Greek: *Haima*, "blood", and *Proteus*, a sea god who had the power of assuming different shapes. The name *Haemoproteus* was first used in the description of *Haemoproteus columbae* in the blood of the pigeon *Columba livia* by Kruse in 1890. This was also the first description of this genus. Two other genera – *Halteridium* and *Simondia* – are now considered to be synonyms of *Haemoproteus*.

The protozoa are intracellular parasites that infect the erythrocytes. They are transmitted by blood sucking insects including mosquitoes, biting midges (*Culicoides*), louse flies (*Hippoboscidae*) and tabanid flies (*Tabanidae*). Infection with this genus is sometimes known as pseudomalaria because of the parasites' similarities with *Plasmodium* species.

Within the genus there are at least 173 species, 5 varieties and 1 subspecies. Of these over 140 occur in birds, 16 in reptiles and 3 in amphibia: 14 orders and 50 families of birds are represented. These include gamebirds (*Galliformes*), waterfowl (*Anseriformes*), raptors (*Accipitriformes, Falconiformes, Strigiformes*), pigeons and doves (*Columbiformes*), and perching birds or songbirds (*Passeriformes*).

### History

The first description of this genus was in 1890 by Kruse who described *Haemoproteus columbae* in the blood of the pigeon *Columba livia*. McCallum in 1897 showed that the process of exflagellation was part of sexual reproduction in these parasites and thought it probable that the same process occurred in *Plasmodium*. The first record of a haemoproteid parasite in a reptile was by Simond in 1901 who gave it the name *Haemamoeba metchnikovi*. The Sergent brothers in 1906 showed that the ectoparasitic fly *Pseudolynchia canariensis* was the vector of *Haemoproteus columbae*. Aragao in 1908 demonstrated the schizogonic stages of *Haemoproteus columbae* in the endothelial cells of the lungs of nestling pigeons infected by the bite of infected *Pseudolynchia*. It was generally believed that transmission of the parasites was by regurgitation during a blood meal until Adie showed that the parasites develop in the salivary glands in a fashion analogous to that of *Plasmodium* in mosquitoes.

The genus *Halterium* was created by the French parasitologist Alphonse Labbe for a species he observed with gametocytes in erythrocytes, with pigment granules, and halter-shaped when fully formed. This genus was soon subsumed into the genus *Haemoproteus*.

The genus *Haemocystidium* was created to give a name to the haemoproteid of a gecko belonging to the genus *Hemidactylus* in Sri Lanka by Castellani and Willey in 1904. A second species in this genus was described in 1909 by Johnston and Cleland who found pigmented gametocytes in the blood of the Australian tortoise *Chelodina longicollis*. These species were transferred to *Haemoproteus* in 1926 by Wenyon.

The genus was resurrected by Garnham in 1966 when he created a new generic name – *Simondia* – for the haemoproteids of chelonians. He followed the opinions of Wenyon, Hewitt and DeGiusti and suggested that all these parasites belonged to the one species – *Simondia metchnikovi*. He retained the name *Haemocystidium* for the haemoproteids of lizards.

A different genus of vectors was identified in 1957 by Fallis and Wood when they identified *Haemoproreus nettionis* in *Culicoides downesi* Wirth and Hubert in Ontario, Canada.

Levine and Campbell in 1971 moved all the species in *Simondia* and *Haemocystidium* into *Haemoproteus* an opinion that was followed by subsequent authors.

The genus *Haemocystidium* was resurrected again by Telford in 1996 when he described three new species of protozoa in geckos from Pakistan.

This genus like those of many protozoa may be further modified once additional DNA sequences are available. For instance, many DNA sequences have been identified for Haemoproteus in birds around the world in recent years, leading to new knowledge about the previously unknown diversity of this parasite in different regions

## Life Cycle

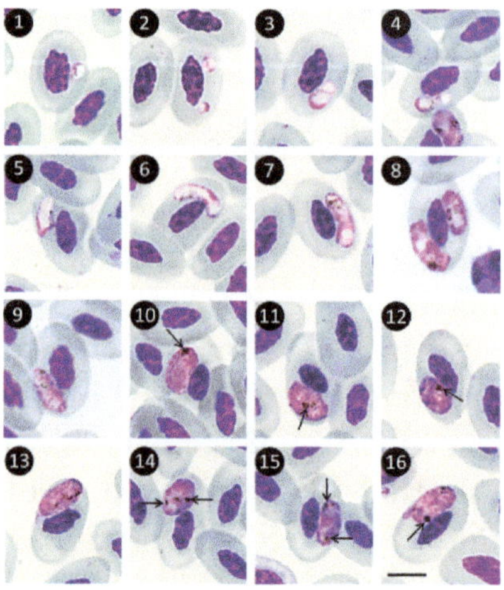


Haemoproteus ilanpapernai

The infective stage is the sporozoite which is present in the salivary glands of the vector. Once the vector bites a new host, the sporozoites enter the blood stream and invade endothelial cells of blood vessels within various tissues including those of the lung, liver and spleen. Within the endothelial cells, the sporozoites undergo asexual reproduction becoming schizonts. These in turn produce numerous merozoites which penetrate the erythrocytes and mature into either female gametocytes (macrogametocytes) or male gametocytes (microgametocytes). Gametocytes can then be ingested by another blood-sucking insect where they undergo sexual reproduction in the midgut of the insect to produce oocysts. The oocysts rupture and release numerous sporozoites that invade the salivary gland and serve as a focus of subsequent infection for another host once the insect takes its next blood meal.

## Evolution

The earliest known fossil is of a *Haemoproteus* like organism (*Paleohaemoproteus burmacis*) was found in the abdominal cavity of a female biting midge trapped 100 million years ago in amber found in Myanmar.

## Diagnostic Criteria

- Gametocytes are only present within erythrocytes
- Gametocytes have a "halter-shaped" appearance with little displacement of the host nucleus
- Schizonts are not seen on peripheral blood smears
- Multiple pigment granules (hemozoin) are present within the erythrocytes

Pigment granules are refractile and yellow to brown in colour.

## General Description

Only gametocytes are found in the blood.

Asexual reproduction occurs in body organs especially the liver.

The organisms occupy the majority of the cytoplasm, leaving the a light magenta, finely granular, pink nucleus centrally located.

Taxonomy of this genus is difficult as there are few distinct morphological differences between the recognised species. Many of them were described under the 'one species-one hos't hypothesis which is now thought to be potentially misleading. The morphological features most commonly used to describe a species include the number of pigment granules, the degree of encirclement of the host nucleus, the size of the parasite, the degree of host nucleus displacement and the degree of host cell enlargement. DNA studies should help to clarify this area but to date have rarely been undertaken.

The gametocytes have five basic forms

- thin gametocytes with incomplete margins (*H. balearicae, H. pelouri*)
- halterial gametocytes (*H. maccullumi*)
- thick sausage shaped gametocytes that fill most of the host cell and displace the host nucleus laterally (*H. halyconis, H. plataleae*)
- gametocytes that encircle the host nucleus and fill the host cell (*H. telfordi*)
- straight gametocytes that normally occur in anucleate cells and are almost as long as the host cell (*H. enucleator*)

## Subgenera

The species infecting avian hosts have been divided into two subgenera – *Haemoproteus* and *Parahaemoproteus* – a division proposed in 1965 by Bennett et al. These may be distinguished as follows:

*Haemoproteus*: Vectors are hippoboscid flies (*Hippoboscidae*). Exflagellation does not occur below 20 degrees Celsius. Mature oocysts have diameters greater than 20 micrometres. The average length of the sporozoites is less than 10 micrometres. One end of the sporozoite is more pointed than the other. Although the majority are parasites of the Columbiformes, some species from this subgenus have also been reported in the Charadriiformes, Pelecaniformes and Suliformes.

*Parahaemoproteus*: Parasites of birds other than the Columbiformes. Vectors are biting midges (*Ceratopogonidae*). Exflagellation occurs below 20 degrees Celsius. Mature oocysts have diameters less than 20 micrometres. The average length of the sporozoites is greater than 10 micrometres. Both ends of the sporozoite are equally pointed.

While it was thought that *Haemoproteus* was limited to doves and related species, species in this genus have been isolated from frigatebirds.

## Pathology

Infections with most *Haemoproteus* species appear to produce subclinical infections.

Post-mortem findings include enlargement of the spleen, liver and kidneys. These organs may appear chocolate-brown due to hemozoin deposition. Cytologic imprints may reveal schizont-laden endothelial cells. Some species of *Haemoproteus* will also form large, cyst-like bodies within the skeletal muscles that resembling those seen with *Sarcocystis* species infections.

Pigeons infected with *Haemoproteus columbae* may develop enlarged gizzards.

Flocks of bobwhite quail (*Colinus virginianus*) may become infected with *Haemoproteus lophortyx*. Infected birds may suffer from reluctance to move, ruffled appearance, prostration and death. Other findings include parasitemia and anemia. Large megaloschizonts may be present in skeletal muscles, particularly those of the thighs and back. The average cumulative mortality for flocks experiencing outbreaks may be over 20%.

Experimental infection of turkeys with *Haemoproteus meleagridis* resulted in lameness, diarrhea, depression, emaciation, anorexia and occasionally anemia.

Muscovey ducks infected with *Haemoproteus nettionis* suffered lameness, dyspnea and sudden death.

In other avian species, anemia and anorexia have been reported occasionally. Importantly, new records of Haemoproteus are discovered constantly and should still be monitored for effects on host condition

## Effect on Vectors

*H. columbae* infects rock pigeons (*Columba livia*) and is vectored by a hippoboscid fly (*Pseudolynchia canariensis*). Both sexes of vector can transmit the parasite. Species of the *Hippoboscoidea* the superfamily to which *Ps. canariensis* belongs do not lay eggs. Instead the larvae hatch *in utero*, are fed internally by 'milk glands' and pass through three morphological stages before being deposited to pupate. The survival of female flies is significantly reduced when they were infected with the parasite. In contrast no effect is seen on male fly survival. Additionally the females produce fewer offspring when infected but the quality of the offspring does not seem to be affected.

## Host Records

## Avian Hosts

- *H. anthi* – yellow wagtails (*Motacilla flava*)
- *H. antigonis* – Florida sandhill crane (*Grus canadensis pratensis*)
- *H. balearicae* – black crowned crane (*Balearica pavonina gibbericeps, Balearica pavonina pavonina*), Florida sandhill crane (*Grus canadensis pratensis*)
- *H. bambusicolae* – bamboo partridge (*Bambusicola thoracica sonorivox*)
- *H. beckeri* – gray catbird (*Dumetella carolinensis*)
- *H. (Parahaemoproteus) belopolskyi* – blackcaps (*Sylvia atricapilla*)
- *H. bennetti* – greater yellownape (*Picus flavinucha*)
- *H. borgesi* – red cockaded woodpecker (*Picoides borealis*)

- *H. brachiatus* – saker falcon (*Falco cherrug*)
- *H. bucerotis* – red billed hornbill (*Tockus erythrorhynchus*)
- *H. (Parahaemoproteus) canachites* – grouse
- *H. (Parahaemoproteus) catharti* – turkey vulture (*Cathartes aura*)
- *H. (Parahaemoproteus) coatneyi* – bananaquit (*Coereba flaveola*), white-crowned sparrows (*Zonotrichia leucophrys*)
- *H. (Haemoproteus) columbae* – Japanese black wood pigeons (*Columba janthina*), pigeon (*Columba livia*), doves (*Columbina talpacoti, Scardafella squammata, Zenaida auriculata*), laughing dove (*Stigmatopelia senegalensis*), eastern white-winged doves (*Zenaida asiatica asiatica*), mourning doves (*Zenaida macroura*)
- *H. (Parahaemoproteus) concavocentralis* - hawfinch (*Coccothraustes coccothraustes*)
- *H. cornuata* – coppersmith barbet (*Megalaima haemacephala*)
- *H. crumenium* – wood stork (*Mycteria americana*)
- *H. (Parahaemoproteus) cyanomitrae* – olive sunbird (*Cyanomitra olivacea*)
- *H. (Parahaemoproteus) danilewskyi* – blue jays (*Cyanocitta cristata*)
- *H. (Parahaemoproteus) desseri* – blossom headed parakeet (*Psittacula roseata*)
- *H. dicruri* – fork tailed drongo (*Dicrurus adsimilis*), crested drongos (*Dicrurus forficatus*)
- *H. elani* – Cooper's hawk (*Accipiter cooperii*), sharp shinned hawk (*Accipiter striatus*)
- *H. enucleator* – kingfisher (*Ispidina picta*)
- *H. (Parahaemoproteus) fringillae* – rufous-winged (*Aimophila carpalis*), house finch (*Carpodacus mexicanus*), hawfinch (*Coccothraustes coccothraustes*), oriental magpie robin (*Copsychus saularis*), dark-eyed juncos (*Junco hyemalis*), American redstarts (*Setophaga ruticilla*)
- *H. forresteri* – rufous-headed ground-roller (*Atelornis crossleyi*)
- *H. gabaldoni* – Muscovy duck (*Cairina moschata*)
- *H. (Parahaemoproteus) garnhami* – sparrows

- *H. goodmani* – pitta-like ground-roller (*Atelornis pittoides*)
- *H. greineri* – wood ducks (*Aix sponsa*), common mergansers (*Mergus merganser*)
- *H. handai* – lesser sulphur-crested cockatoo (*Cacatua sulphurea*), plum-headed parakeet (*Psittacula cyanocephala*), ring necked parakeet (*Psittacula krameri manillensis*)
- *H. himalayanus* – rufous sibia (*Heterophasia capistrata*)
- *H. (Parahaemoproteus) homobelopolskyi* - red headed malimbe (*Malimbus rubricollis*), black headed weaver (*Ploceus melanocephalus*), red billed quelea (*Quelea quelea*)
- *H. (Parahaemoproteus) homovelans* - grey-faced woodpecker (*Picus canus*)
- *H. ilanpapernai* - spotted wood owl (*Strix seloputo*), Brown Hawk-Owl (*Ninox scutulata*)
- *H. iwa* – great frigatebirds (*Fregata minor*), magnificent frigatebirds (*Fregata magnificens*)
- *H. janovyi* – whitebacked vulture (*Gyps africanus*), hooded vulture (*Necrosyrtes monachus*), white-headed vulture (*Trigonoceps occipitalis*) lappet faced vulture (*Torgos tracheliotus*)
- *H. (Haemoproteus) jenniae* - swallow tailed gull (*Creagrus furcatus*)
- *H. khani* – crested drongos (*Dicrurus forficatus*)
- *H. (Parahaemoproteus) lanii* – red backed shrike (*Lanius collurio*), woodchat shrike (*Lanius senator*)
- *H. lari* – Caspian gulls (*Larus cachinnans*)
- *H. (Parahaemoproteus) lophortyx* – California quail (*Callipepla californica*), scaled quail (*Callipepla squamata*), bobwhite quail (*Colinus virginianus*)
- *H. maccallumi* – mourning doves (*Zenaida macroura*)
- *H. macrovacuolatus* - black-bellied whistling duck (*Dendrocygna autumnalis*)
- *H. madagascariensis* – hook billed vanga (*Vanga curvirostris*)
- *H. majoris* – Swainson's thrush (*Catharus ustulatus*), blue tits (*Cyanistes caeruleus, Parus caeruleus*)
- *H. mansoni* – blue grouse (*Dendragapus obscurus*), ptarmigan (*Lagopus lagopus*)

- *H. meleagridis* – turkey (*Meleagris gallopavo*)
- *H. (Parahaemoproteus) micronuclearis* - red headed malimbe (*Malimbus rubricollis*), black headed weaver (*Ploceus melanocephalus*), red billed quelea (*Quelea quelea*)
- *H. (Haemoproteus) multipigmentatus* – Galapagos dove (*Zenaida galapagoensis*)
- *H. motacillae* – yellow wagtails (*Motacilla flava*)
- *H. (Haemoproteus) multivolutinus* - tambourine dove (*Turtur timpanistria*)
- *H. (Parahaemoproteus) nettionis* – wood ducks (*Aix sponsa*), blue-winged teals (*Anas discors*), Pekin duck (*Anas platyrhynchos*), lesser scaups (*Aythya affinis*), ring-necked ducks (*Aythya collaris*), Muscovey duck (*Cairina moschata*), trumpeter swans (*Cygnus buccinator*)
- *H. nisi* – Cooper's hawk (*Accipiter cooperii*), sharp shinned hawk (*Accipiter striatus*)
- *H. (Parahaemoproteus) nucleofascialis* - red headed malimbe (*Malimbus rubricollis*), black headed weaver (*Ploceus melanocephalus*), red billed quelea (*Quelea quelea*)
- *H. noctuae* – snowy owls (*Nyctea scandiaca*), spotted owl (*Strix occidentalis*)
- *H. orioli* – golden oriole (*Oriolus oriolus*)
- *H. oryzivorae* – oriental magpie robin (*Copsychus saularis*), Indian silverbill (*Lonchura malabarica*), tricoloured munia (*Lonchura malacca ruboniger*), scaly-breasted munia (*Lonchura punctulata*), baya weaver (*Ploceus philippinus*), jungle babbler (*Turdoides striata*)
- *H. palumbus* – pigeon (*Columba palumbus palumbus*)
- *H. pallidulus* – blackcap (*Sylvia atricapilla*)
- *H. parabelopolskyi* – blackcap (*Sylvia atricapilla*)
- *H. (Haemoproteus) paramultipigmentatus* - Socorro common ground dove (*Columbina passerina socorroensis*)
- *H. passeris* – Israeli house sparrow (*Passer domesticus biblicus*)
- *H. pasteris* – pied myna (*Sturnus contra*), grey headed myna (*Sturnus malabaricus*)
- *H. pastoris* – greater blue eared glossy starling (*Lamprotornis chalybaeus*), pied myna (*Sturnus contra*)

- *H. payevskyi* – great reed warbler (*Acrocephalus arundinaceus*), marsh warbler (*Acrocephalus palustris*)
- *H. (Haemoproteus) piresi* – pigeon (*Columba livia*)
- *H. plataleae* – glossy ibis (*Plegadis falcinellus*)
- *H. pratosi* – Ahanta francolin (*Francolinus ahantensis*)
- *H. pratasi* – helmeted guineafowl (*Numida meleagris*)
- *H. prognei* – purple martin (*Progne subis*)
- *H. psittaci* – African grey parrot (*Psittacus erithacus*)
- *H. raymundi* – eastern olive sunbird (*Nectarinia olivacea*)
- *H. (Haemoproteus) sacharovi* – eastern white-winged doves (*Zenaida asiatica asiatica*), mourning doves (*Zenaida macroura*)
- *H. sangunis* – red whiskered bulbul (*Pycnonotus jocosus emeria*)
- *H. (Parahaemoproteus) sanîosdiasï* – chicken (*Gallus gallus*)
- *H. silvaï* – guinea fowl (*Numida meleagris mitrata*)
- *H. sylvae* – great reed warbler (*Acrocephalus arundinaceus*)
- *H. syrnii* – tawny owl (*Strix aluco*), spotted owl (*Strix occidentalis*), European scops owl (*Otus scops*)
- *H. telfordi* – MacQueen's bustards (*Chlamydotis macqueenii*), rufous-crested bustards (*Eupodotis ruficrista*), great bustard (*Otis tarda*)
- *H. tendeiroi* – MacQueen's bustards (*Chlamydotis macqueenii*), rufous-crested bustards (*Eupodotis ruficrista*), great bustard (*Otis tarda*)
- *H. tinnunculi* – American kestrel (*Falco sparverius*), Chimango caracara (*Milvago chimango*)
- *H. (Haemoproteus) turtur* – turtle dove (*Streptopelia turtur*)
- *H. (Parahaemoproteus) vacuolatus* – yellow whiskered greenbul (*Andropadus latirostris*)
- *H. (Parahaemoproteus) valkiūnasi* – great frigatebirds (*Fregata minor*), lesser frigatebirds (*Fregata ariel*), Ascension frigatebirds (*Fregata aquila*)
- *H. vangii* – hook billed vanga (*Vanga curvirostris*)

- *H. (Parahaemoproteus) velans* – red-bellied woodpecker (*Melanerpes carolinus*), red-cockaded woodpecker (*Picoides borealis*)
- *H. zosteropsis* – oriental white eye (*Zosterops palpebrosa palpebrosa*)

## Reptile Hosts

- *H. anatolicum* - tortoise (*Testudo graeca*)
- *H. balli* – Egyptian cobra (*Naja haje haje*)
- *H. chelodina* – saw-shelled tortoise (*Elseya latisternum*)
- *H. edomensis* – lizard (*Agama stellio*)
- *H. geochelonis* – tortoise (*Geochelone denticulata*)
- *H. kopki* – spotted Indian house gecko (*Hemidactylus brookei*), giant frog eye gecko (*Teratoscincus scincus*)
- *H. mackerrasi* – Binoe's prickly gecko (*Heteronotia binoei*)
- *H. mesnili* – spitting cobra (*Naja nigricollis nigricolli*)
- *H. metchnikovi* – turtle (*Chrysemys picta*), yellow bellied terrapin (*Tramchemys scripta*)
- *H. oedurae* – Australian northern velvet gecko (*Oedura castelnaui*)
- *H. peltocephali* – river turtle (*Peltocephalus dumerilianus*)
- *H. phyllodactyli* – gekkonid (*Ptyodactylus elisa*)
- *H. ptyodactyli* – Kramer's yellow fan-fingered gecko (*Ptyodactylus hasselquistii*)
- *H. tarentolae* – Moorish gecko (*Tarentola mauritanica*)
- *H. trionyxi* – Ganges softshell turtle (*Trionyx gangeticus*)

## Amphibian Hosts

- *H. ovalis* – cricket frog (*Rana limnocharis*)

## Hosts known to be Infected but Haemoproteus Species not Identified

- common myna (*Acridotheres tristis*)
- Blyth's reed warbler (*Acrocephalus dumetorum*)

- sedge warblers (*Acrocephalus schoenobaenus*)
- reed warbler (*Acrocephalus scirpaceus*)
- clamorous reed warbler (*Acrocephalus stentoreus*)
- black throated sunbird (*Aethopyga saturata*)
- Spanish red-legged partridge (*Alectoris rufa*)
- imperial eagles (*Aquila heliaca*)
- canvasbacks (*Aythya valisineria*)
- white cockatoo (*Cacatua alba*)
- sulphur-crested cockatoo (*Cacatua galerita*)
- speckled pigeon (*Columba guinea*)
- white-rumped shama (*Copsychus malabaricus*)
- green jays (*Cyanocorax yncas glaucescens*)
- European bee-eaters (*Merops apiaster*)
- mute swan (*Cygnus olor*)
- magnificent bird of paradise (*Diphyllodes magnificus hunsteini*)
- red munia (*Estrilda amandava*)
- lesser kestrel (*Falco naumanni*)
- common kestrel (*Falco tinnunculus*)
- Swainson's francolin (*Francolinus swainsonii*)
- magnificent frigatebirds (*Fregata magnificens*)
- chaffinch (*Fringilla coelebs*)
- hill mynah (*Gracula religiosa intermedia*)
- long tailed shrike (*Lanius schach*)
- superb bird of paradise (*Lophorina superba*)
- Egyptian kites (*Milvus migrans aegypticus*)
- Guianan red-capped cardinal (*Paroaria gularis gularis*)

- lesser flamingos (*Phoeniconaias minor*)
- New Holland honeyeaters (*Phylidonyris novaehollandiae*)
- streaked weaver (*Ploceus manyar*)
- Surinam crested oropendola (*Psarocolius decumanus decumanus*)
- Montezuma oropendolas (*Psarocolius montezuma*)
- Guianan turquoise tanager (*Tangara mexicana mexicana*)
- blue-necked tanager (*Tangara cyanicollis caeruleocephala*)
- sacred ibis (*Threskiornis aethiopicus*)
- white-crowned sparrows (*Zonotrichia leucophrys oriantha*)

## Vectors

- *H. balmorali – Culicoides impunctatus*
- *H. belopolskyi – Culicoides impunctatus*
- *H. columbae – Ornithomyia avicularia, Pseudolynchia canariensis*
- *H. danilewskyi – Culicoides arboricola, Culicoides edeni, Culicoides knowltoni*
- *H. dolniki – Culicoides impunctatus*
- *H. fringillae – Culicoides impunctatus*
- *H. lanii – Culicoides impunctatus*
- *H. lophortyx – Culicoides bottimeri, Lynchia hirsuta, Stilbometopa impressa*
- *H. metchinikovi – Chrysops callidus*
- *H. nettionis – Culicoides downesi*
- *H. sacharovi – Peseudolynchia maura*
- *H. syrnii – Ornithomyia avicularia*
- *H. tartakovskyi – Culicoides impunctatus*
- *H. turtur – Pseudolynchia canariensis*

## Avian Families Affected

The concept of a "one host-one species" was originally used in the taxonomy of this genus as it appears that the parasites are at least moderately host specific. After this rule was found to be incorrect, it was suggested that the avian parasite species were limited to single avian families. From an inspection of the host records above it is clear that this is not the case.

The avian species known to be infected are listed below:

Order Accipitriformes

Family Accipitridae

- Cooper's hawk (*Accipiter cooperii*)
- sharp shinned hawk (*Accipiter striatus*)
- eastern imperial eagle (*Aquila heliaca*)
- white-backed vulture (*Gyps africanus*)
- black kite (*Milvus migrans*)
- hooded vulture (*Necrosyrtes monachus*)
- white-headed vulture (*Trigonoceps occipitalis*)
- lappet faced vulture (*Torgos tracheliotos*)

Family Cathartidae

- Turkey vulture (*Cathartes aura*)

Order Anseriformes

Family Anatidae

- wood duck (*Aix sponsa*)
- blue winged teal (*Anas discors*)
- mallard duck (*Anas platyrhynchos*)
- lesser scaup (*Aythya affinis*)
- ring necked duck (*Aythya collaris*)
- canvasback (*Aythya valisineria*)
- Muscovy duck (*Cairina moschata*)
- trumpeter swan (*Cygnus buccinator*)

- mute swan (*Cygnus olor*)
- black-bellied whistling duck (*Dendrocygna autumnalis*)
- common merganser (*Mergus merganser*)

Order Charadriiformes

Family Laridae

- swallow tailed gull (*Creagrus furcatus*)
- Caspian gull (*Larus cachinnans*)

Order Ciconiiformes

Family Ciconiidae

- wood stork (*Mycteria americana*)

Order Columbiformes

Family Columbidae

- speckled pigeon (*Columba guinea*)
- Japanese wood pigeon (*Columba janthina*)
- rock pigeon (*Columba livia*)
- common wood pigeon (*Columba palumbus*)
- Socorro common ground dove (*Columbina passerina socorroensis*)
- ruddy ground dove (*Columbina talpacoti*)
- tambourine dove (*Turtur timpanistria*)
- scaled dove (*Scardafella squammata*)
- European turtle dove (*Streptopelia turtur*)
- laughing dove (*Stigmatopelia senegalensis*)
- white-winged dove (*Zenaida asiatica*)
- eared dove (*Zenaida auriculata*)
- Galápagos Dove (*Zenaida galapagoensis*)
- mourning dove (*Zenaida macroura*)

Order Coraciiformes

Family Alcedinidae

- African pygmy kingfisher (*Ispidina picta*)

Family Brachypteraciidae

- rufous headed ground roller (*Atelornis crossleyi*)
- pitta like ground roller (*Atelornis pittoides*)

Family Bucerotidae

- red-billed hornbill (*Tockus erythrorhynchus*)

Order Falconiformes

Family Falconidae

- saker falcon (*Falco cherrug*)
- lesser kestrel (*Falco naumanni*)
- American kestrel (*Falco sparverius*)
- common kestrel (*Falco tinnunculus*)
- chimango caracara (*Milvago chimango*)

Order Galliformes

Family Numididae

- helmeted guineafowl (*Numida meleagris*)

Family Odontophoridae

- California quail (*Callipepla californica*)
- scaled quail (*Callipepla squamata*)
- bobwhite quail (*Colinus virginianus*)

Family Phasianidae

- red legged partridge (*Alectoris rufa*)
- Chinese bamboo partridge (*Bambusicola thoracicus*)
- Ahanta francolin (*Francolinus ahantensis*)

- Swainson's francolin (*Francolinus swainsonii*)
- chicken (*Gallus gallus*)
- willow grouse (*Lagopus lagopus*)
- wild turkey (*Meleagris gallopavo*)

Family Tetraonidae

- dusky grouse (*Dendragapus obscurus*

Order Gruiformes

Family Gruidae

- black crowned crane (*Balearica pavonina*)
- sandhill crane (*Grus canadensis*)

Family Otidae

- MacQueen's bustard (*Chlamydotis macqueenii*)
- red crested bustard (*Eupodotis ruficrista*)
- great bustard (*Otis tarda*)

Order Passeriformes

Family Acrocephalidae

- great reed warbler (*Acrocephalus arundinaceus*)
- Blyth's reed warbler (*Acrocephalus dumetorum*)
- marsh warbler (*Acrocephalus palustris*)
- sedge warbler (*Acrocephalus schoenobaenus*)
- reed warbler (*Acrocephalus scirpaceus*)
- clamorous reed warbler (*Acrocephalus stentoreus*)

Family Corvidae

- blue jay (*Cyanocitta cristata*)
- green jay (*Cyanocorax yncas*)

Family Dicruridae

- fork-tailed drongo (*Dicrurus adsimilis*)
- crested drongo (*Dicrurus forficatus*)

Family Emberizidae

- dark eyed junco (*Junco hyemalis*)
- rufous winged sparrow (*Peucaea carpalis*)
- white crowned sparrow (*Zonotrichia leucophrys*)

Family Estrildidae

- red munia (*Amandava amandava*)
- Indian silverbill (*Euodice malabarica*)
- tricoloured munia (*Lonchura malacca*)
- scaly breasted munia (*Lonchura punctulata*)

Family Fringillidae

- house finch (*Carpodacus mexicanus*)
- hawfinch (*Coccothraustes coccothraustes*)
- chaffinch (*Fringilla coelebs*)

Family Hirundinidae

- purple martin (*Progne subis*)

Family Icteridae

- crested oropendola (*Psarocolius decumanus*)
- Montezuma oropendola (*Psarocolius montezuma*)

Family Laniidae

- red backed shrike (*Lanius collurio*)
- long tailed shrike (*Lanius schach*)
- woodchat shrike (*Lanius senator*)

Family Meliphagidae

- New Holland honeyeaters (*Phylidonyris novaehollandiae*)

Family Mimidae

- gray catbird (*Dumetella carolinensis*)

Family Motacillidae

- yellow wagtail (*Motacilla flava*)

Family Muscicapidae

- white rumped shama (*Copsychus malabaricus*)
- oriental magpie robin (*Copsychus saularis*)

Family Nectariniidae

- black throated sunbird (*Aethopyga saturata*)
- olive sunbird (Cyanomitra olivacea)

Family Oriolidae

- golden oriole (*Oriolus oriolus*)

Family Paridae

- blue tit (*Cyanistes caeruleus*)

Family Paradisaeidae

- magnificent bird of paradise (*Cicinnurus magnificus*)
- superb bird of paradise (*Lophorina superba*)

Family Parulidae

- American redstart (*Setophaga ruticilla*)

Family Passeridae

- house sparrow (*Passer domesticus*)

Family Ploceidae

- red headed malimbe (*Malimbus rubricollis*)
- streaked weaver (*Ploceus manyar*)
- black-headed weaver (*Ploceus melanocephalus*)
- baya weaver (*Ploceus philippinus*)

- red billed quelea (*Quelea quelea*)

Family Pycnonotidae

- yellow whiskered bulbul (*Andropadus latirostris*)
- red whiskered bulbul (*Pycnonotus jocosus*)

Family Sturnidae

- common myna (*Acridotheres tristis*)
- common hill myna (*Gracula religiosa*)
- chestnut tailed starling (*Sturnia malabarica*)
- Asian pied starling (*Sturnus contra*)
- greater blue eared glossy starling (*Lamprotornis chalybaeus*)

Family Sylviidae

- blackcap (*Sylvia atricapilla*)

Family Thraupidae

- bananaquit (*Coereba flaveola*)
- red capped cardinal (*Paroaria gularis*)
- blue necked tanager (*Tangara cyanicollis*)
- turquoise tanager (*Tangara mexicana*)

Family Timaliidae

- rufous sibia (*Heterophasia capistrata*)
- jungle babbler (*Turdoides striata*)

Family Turdidae

- Swainson's thrush (*Catharus ustulatus*)

Family Vangidae

- hook billed vanga (*Vanga curvirostris*)

Family Zosteropidae

- Oriental white eye (*Zosterops palpebrosus*)

Order Pelecaniformes

Family Fregatidae

- magnificent frigatebird (*Fregata magnificens*)
- great frigatebird (*Fregata minor*)

Family Threskiornithidae

- African sacred ibis (*Threskiornis aethiopicus*)
- glossy ibis (*Plegadis falcinellus*)

Order Piciformes

Family Megalaimidae

- coppersmith barbet (*Megalaima haemacephala*)

Family Picidae

- red bellied woodpecker (*Melanerpes carolinus*)
- red cockaded woodpecker (*Picoides borealis*)
- grey faced woodpecker (*Picus canus*)
- greater yellownape (*Picus flavinucha*)

Order Phoenicopteriformes

Family Phoenicopteridae

- lesser flamingo (*Phoenicopterus minor*)

Order Psittaciformes

Family Cacatuidae

- white cockatoo (*Cacatua alba*)
- sulphur crested cockatoo (*Cacatua galerita*)
- yellow crested cockatoo (*Cacatua sulphurea*)

Family Psittacidae

- plum headed parakeet (*Psittacula cyanocephala*)
- rose ringed parakeet (*Psittacula krameri*)

- blossom headed parakeet (*Psittacula roseata*)
- African grey parrot (*Psittacus erithacus*)

Order Strigiformes

Family Strigidae

- snowy owl (*Bubo scandiacus*)
- Brown Hawk-Owl (*Ninox scutulata*)
- European scops owl (*Otus scops*)
- brown owl (*Strix aluco*)
- spotted owl (*Strix occidentalis*)
- spotted wood owl (*Strix seloputo*)

### Notes

*Haemoproteus balazuci* Dias 1953 is a junior synonym of *Haemoproteus testudinalis*

*Haemoproteus gymnorhidis* de Mello 1936, *Haemoproteus granulosum* Rey Vila 1945, *Haemoproteus danilewskyi var. urbanensis* Sachs 1953 and *Haemoproteus zasukhini* Burtikashvili 1973 are considered to be synonyms of *Haemoproteus passeris* Kruse 1890.

*Haemoproteus rouxi* Novy and MacNeal 1904 is a *nomen nudum*.

## Marek's Disease

Marek's disease is a highly contagious viral neoplastic disease in chickens. It is named after József Marek, a Hungarian veterinarian. Occasionally misdiagnosed as an ab-tissue pathology it is caused by an alphaherpesvirus known as 'Marek's disease virus' (MDV) or *Gallid herpesvirus 2* (GaHV-2). The disease is characterized by the presence of T cell lymphoma as well as infiltration of nerves and organs by lymphocytes. Viruses related to MDV appear to be benign and can be used as vaccine strains to prevent Marek's disease. For example, the related Herpesvirus of Turkeys (HVT), causes no apparent disease in turkeys and continues to be used as a vaccine strain for prevention of Marek's disease. Birds infected with GaHV-2 can be carriers and shedders of the virus for life. Newborn chicks are protected by maternal antibodies for a few weeks. After infection, microscopic lesions are present after one to two weeks, and gross lesions are present after three to four weeks. The virus is spread in dander from feather follicles and transmitted by inhalation.

## Syndromes

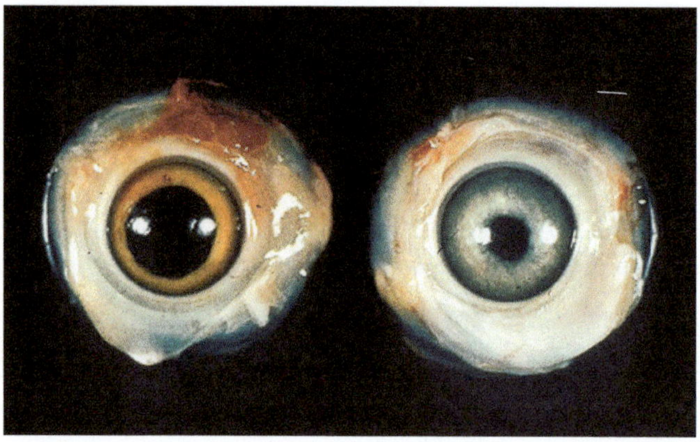

Left—normal chicken eye. Right—Eye of a chicken with Marek's disease

Six syndromes are known to occur after infection with Marek's disease. These syndromes may overlap.

- Classical Marek's disease or neurolymphomatosis causes asymmetric paralysis of one or more limbs. With vagus nerve involvement, difficulty breathing or dilation of the crop may occur. Besides lesions in the peripheral nerves, there are frequently lymphomatous infiltration/tumours in the skin, skeletal muscle, visceral organs. Organs that are commonly affected include the ovary, spleen, liver, kidneys, lungs, heart, proventriculus and adrenals.
- Acute Marek's disease is an epidemic in a previously uninfected or unvaccinated flock, causing depression, paralysis, and death in a large number of birds (up to 80%). The age of onset is much earlier than the classic form; birds are four to eight weeks old when affected. Infiltration into multiple organs/tissue is observed.
- Ocular lymphomatosis causes lymphocyte infiltration of the iris (making the iris turn grey), unequal size of the pupils, and blindness.
- Cutaneous Marek's disease causes round, firm lesions at the feather follicles.
- Atherosclerosis is induced in experimentally infected chickens.
- Immunosuppression – Impairment of the T-lymphocytes prevents competent immunological response against pathogenic challenge and the affected birds become more susceptible to disease conditions such as coccidiosis and *Escherichia coli* infection . Furthermore, without stimulation by cell-mediated immunity, the humoral immunity conferred by the B-cell lines from the Bursa of Fabricius also shuts down, thus resulting in birds that are totally immunocompromised.

## Diagnosis

Diagnosis of lymphoid tumors in poultry is complicated due to multiple etiological agents capable of causing very similar tumors. It is not uncommon that more than one avian tumor virus can be present in a chicken, thus one must consider both the diagnosis of the disease/tumors (pathological diagnosis) and of the virus (etiological diagnosis). A step-wise process has been proposed for diagnosis of Marek's disease which includes (1) history, epidemiology, clinical observations and gross necropsy, (2) characteristics of the tumor cell, and (3) virological characteristics

The demonstration of peripheral nerve enlargement along with suggestive clinical signs in a bird that is around three to four months old (with or without visceral tumors) is highly suggestive of Marek's disease. Histological examination of nerves reveals infiltration of pleomorphic neoplastic and inflammatory lymphocytes. Peripheral neuropathy should also be considered as a principal rule-out in young chickens with paralysis and nerve enlargement without visceral tumors, especially in nerves with interneuritic edema and infiltration of plasma cells.

The presence of nodules on the internal organs may also suggest Marek's disease, but further testing is required for confirmation. This is done through histological demonstration of lymphomatous infiltration into the affected tissue. A range of leukocytes can be involved, including lymphocytic cell lines such as large lymphocyte, lymphoblast, primitive reticular cells, and occasional plasma cells, as well as macrophage and plasma cells. The T cells are involved in the malignancy, showing neoplastic changes with evidence of mitosis. The lymphomatous infiltrates need to be differentiated from other conditions that affect poultry including lymphoid leukosis and reticuloendotheliosis, as well as an inflammatory event associated with hyperplastic changes of the affected tissue.

Key clinical signs as well as gross and microscopic features that are most useful for differentiating Marek's disease from lymphoid leukosis and reticuloendotheliosis include (1) Age: MD can affect birds at any age, including <16 weeks of age; (2) Clinical signs: Frequent wing and leg paralysis; (3) Incidence: >5% in unvaccinated flocks; (4) Potential nerve enlargement; (5) Interfollicular tumors in the bursa of Fabricius; (6) CNS involvement; (7) Lymphoid proliferation in skin and feather follicles; (8) Pleomorphic lymphoid cells in nerves and tumors; and (9) T-cell lymphomas.

In addition to gross pathology and histology, other advanced procedures used for a definitive diagnosis of Marek's disease include immunohistochemistry to identify cell type and virus-specific antigens, standard and quantitative PCR for identification of the virus, virus isolation to confirm infections, and serology to confirm/exclude infections.

The World Organisation for Animal Health (OIE) reference laboratories for Marek's disease include the Institute for Animal Health, Compton Laboratory, UK and the USDA Avian Disease and Oncology Laboratory, USA.

## Prevention

Vaccination is the only known method to prevent the development of tumors when chickens are infected with the virus. However, administration of vaccines does not prevent transmission of the virus, i.e., the vaccine is not sterilizing. However, it does reduce the amount of virus shed in the dander, hence reduces horizontal spread of the disease. Marek's disease does not spread vertically. The vaccine was introduced in 1970 and the scientist credited with its development is Dr. Ben Roy Burmester. Before that, Marek's disease caused substantial revenue loss in the poultry industries of the United States and the United Kingdom. The vaccine can be administered to one-day-old chicks through subcutaneous inoculation or by *in ovo* vaccination when the eggs are transferred from the incubator to the hatcher. *In ovo* vaccination is the preferred method, as it does not require handling of the chicks and can be done rapidly by automated methods. Immunity develops within two weeks.

The vaccine originally contained the antigenically similar turkey herpesvirus, which is serotype 3 of MDV. However, because vaccination does not prevent infection with the virus, the Marek's disease virus has evolved increased virulence and resistance to this vaccine. As a result, current vaccines use a combination of vaccines consisting of HVT and gallid herpesvirus type 3 or an attenuated MDV strain, CVI988-Rispens (ATCvet code: QI01AD03 (WHO)).

## References


- Cox, N.; Kawaoka (1998). "22". In Mahy B. and Collier L. Topley and Wilson's Microbiology and Microbial Infections. 1 Virology. Y. (9 ed.). Arnold. p. 415. ISBN 0-340-61470-6.
- Franson JC, Friend M. "38: Avian Botulism". Field Manual of Wildlife Disease (PDF). U.S. Geological Survey. ISBN 978-0-607-88096-0. Retrieved 14 June 2016.
- Klaus, Sidney N.; Shoshana Frankenburg, and A. Damian Dhar (2003). "Chapter 235: Leishmaniasis and Other Protozoan Infections". In Freedberg; et al. Fitzpatrick's Dermatology in General Medicine. (6th ed.). McGraw-Hill. ISBN 0-07-138067-1.
- Toxoplasma gondii: the model apicomplexan: perspectives and methods. Elsevier/Academic Press. 2007. ISBN 9780123695420.
- James, William D.; Berger, Timothy G.; et al. (2006). Andrews' Diseases of the Skin: clinical Dermatology. Saunders Elsevier. p. 260. ISBN 0-7216-2921-0.
- Nigar Kirmani, Keith F. Woeltje, Hilary Babcock (2012). The Washington Manual of Infectious Disease Subspecialty Consult. Lippincott Williams & Wilkins. ISBN 9781451113648. Page 194
- Green, Elizabeth Alden (1979). Mary Lyon and Mount Holyoke. Hanover, New Hampshire: University Press of New England. p. 310. ISBN 0-87451-172-0.
- Capaldi, Nicholas (2004). John Stuart Mill: a biography. Cambridge, UK: Cambridge University Press. p. 356. ISBN 0-521-62024-4.
- Wollenweber, Brother Leo (2002). "Meet Solanus Casey". St. Anthony Messenger Press, Cincinnati, Ohio, page 107, ISBN 1-56955-281-9,

- Rice, Christopher (1990). Lenin: Portrait of a Professional Revolutionary. London: Cassell. ISBN 978-0304318148. pp 77-78
- Hirai, K, ed. (2001). Current Topics in Microbiology and Immunology: Marek's Disease (Current Topics in Microbiology and Immunology). Springer: Berlin. ISBN 3-540-67798-4.
- Fenner, Frank J.; Gibbs, E. Paul J.; Murphy, Frederick A.; Rott, Rudolph; Studdert, Michael J.; White, David O. (1993). Veterinary Virology (2nd ed.). Academic Press, Inc. ISBN 0-12-253056-X

8

# Various Poultry Standards

The various poultry standards elucidated in this chapter are American standard of perfection, Australian poultry standards and British poultry standard. These ensure that efficient and effective food safety practices are being maintained. The aspects elucidated in this section are of vital importance, and provides a better understanding of poultry science.

## American Standard of Perfection

The *American Standard of Perfection* is the official breed standard for the poultry fancy in North America. First published in 1874 by the American Poultry Association, the *Standard of Perfection* (commonly referred to as "the *Standard*") classifies and describes the standard physical appearance, coloring and temperament for all recognized breeds of poultry, including chickens, ducks, turkeys, and geese. The current edition was published in 2010.

The cover of the 1930 edition

The *Standard* is used by American Poultry Association judges at sanctioned poultry shows to judge poultry, and by those who participate in the competitive showing of selectively bred birds that conform to the standard, which led to the term "standard bred" poultry.

The first edition of the book listed 41 breeds, and today's versions have nearly 60. There are 19 classes of poultry recognized by the American Poultry Association. Eleven of these classes are devoted to chickens, of which six are classes of large breeds and five are bantam classes. There are four classes of ducks and three classes of geese, both divided by weight. All breeds of turkeys are grouped into one class.

### List of American Poultry Association Classes

Standard: American, Asiatic, Continental, English, Mediterranean, and All Other Standard Breeds

Bantam: Single Comb Clean Legged (SCCL), Rose Comb Clean Legged (RCCL), All Other Comb Clean Legged (AOCCL), Feather Legged, and Game Bantam

Ducks: Heavy, Medium, Light, and Bantam

Geese: Heavy, Medium, and Light

Turkeys: Judged as one class.

Guinea Fowl: Judged as one class.

## Australian Poultry Standards

The cover of the 1998 first edition

The *Australian Poultry Standards* is the official breed standard for poultry fancy in Australia. It is the standard of perfection from which all poultry in Australia is supposed to be judged when exhibited at poultry shows. It is published by Victorian Poultry Fanciers Association, the peak body for poultry in Victoria and agreed to by all other state peak bodies.

## Publication

Published by the Victorian Poultry Fanciers Association Inc., the first edition of the *Australian Poultry Standard* was published in 1998. It has been replaced by the second edition.

Australia has no national peak body for poultry, relying on state bodies and national breed clubs. The state bodies who are party to the standards are as follows (None of the mainland territories have a statewide poultry society)

- Victorian Poultry Fanciers Association (trades as Poultry Stud Breeders and Exhibitors Victoria) (founded 1970)
- Tasmanian Poultry Fanciers Association
- South Australian Poultry Association
- Exhibition Poultry Association of New South Wales
- Western Australian Poultry Association (founded 1889)
- Feather Clubs Association of Queensland (founded 1959)

## Classification of Breeds

The *Australian Poultry Standard* classifies all chicken breeds as either hard feather or soft feather, and then further divides them into the bantam or standard size. Breeds which have less under fluff and feathers which are closer to the body fall under hard feather, such as the Old English Game and Asil. Breeds with more under fluff and softer feathers, such as the Silkie, Australorp and Leghorn, are classed as soft feather.

The *Australian Poultry Standards* also includes standards for ducks, geese, turkeys and Guinea fowl. Ducks are divided into two classes, bantam or standard size, and all turkeys, geese and guinea fowl have only one class each.

## Earlier Publications

In the 1940s and earlier, other publications in Australia about poultry, such as Hadlingtons, did have some parts devoted to aspects of showing birds, were less oriented towards show standards, and more detailed about general management of poultry, in line with his earliest publication.

16 *Hadlington's Australian Poultry Book*

shown to the innovation. To-day, the necessity for such a standard has been almost universally recognized by poultry breeders throughout Australia and elsewhere.

Fig. 2. Points of a fowl.

1. Comb.
2. Face.
3. Wattles.
4. Ear-lobes.
5. Hackle.
6. Breast.
7. Back.
8. Saddle.
9. Saddle hackles.
10. Sickles.
11. Tail coverts.
12. True tail feathers.
13. Wing-bow.
14. Wing-coverts, forming the bar.
15. Secondaries—the lower ends forming the wing or lower butts. Wing bay—diamond.
16. Lower wing butts.
17. Primaries, hidden by secondaries when the wing is closed.
18. Thighs.
19. Hocks.
20. Legs or shanks.
21. Spurs.
22. Toes or claws.

For some years now the novice poultry farmer has been led to believe that by procuring certain lines of blood, deemed to be of high productive strain, and by selecting his breeders on a system of measuring the pelvic bone, abdomen, etc., a knowledge of breed characteristics is unnecessary. Unfortunately only too

Hadlingtons' *Points of Fowl* from the 1940s

## British Poultry Standard

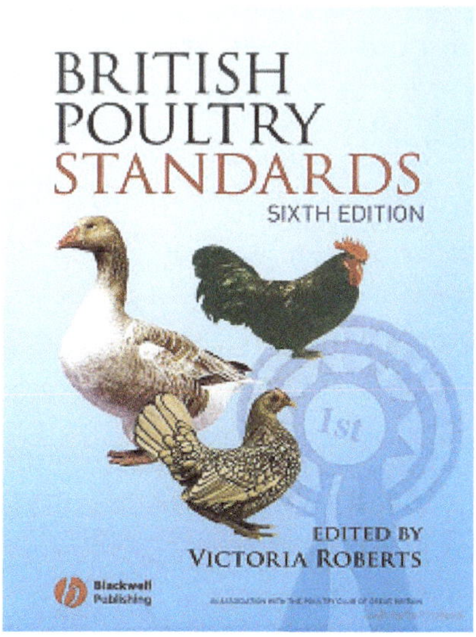


The front cover of the 6th Edition of the British Poultry Standards.

The *British Poultry Standard* is the oldest poultry fancy breed standard in the world. It is published by the Poultry Club of Great Britain and is the official reference standard used by judges at poultry shows within the United Kingdom and the Republic of Ireland.

## History

The standard was first published in 1865 by the original Poultry Club of Great Britain, a club which existed for only three years. It was entitled *The standard of excellence in Exhibition Poultry* and was edited by William Bernhardt Tegetmeier. It was the first publication of its kind. The compilation of the standard was then taken over by the second, current Poultry Club of Great Britain. The number of editions of the standard that have been published is uncertain, as each successive publisher that has been used by the Poultry Club of Great Britain has started again with a first edition. The current edition, published in 2008 by Blackwell, is the sixth in the current numbering.

## Use

The standard is the official reference for use by judges at poultry shows in the United Kingdom and the Republic of Ireland. It contains breed standards of more than 150 breeds of chickens, ducks, geese and turkeys; pigeons are not included.

## Editions

Editions of the standards include, but probably are not limited to, these:

- William Bernhard Tegetmeier (editor). *The Standard of Excellence in Exhibition Poultry, authorized by the Poultry Club*. London: Poultry Club, 1865.
- ——. *The Poultry Book: comprising the breeding and management of profitable and ornamental poultry, their qualities and characteristics; to which is added the Standard of Excellence in exhibition birds, authorized by the Poultry Club*. [S.l.]: Routledge, 1867.
- ——, Harrison Weir (illustrator). *The Poultry Book: comprising the breeding and management of profitable and ornamental poultry, their qualities and characteristics; to which is added "The Standard of Excellence in Exhibition Birds," authorized by the Poultry Club*. London; New York : George Routledge and Sons, the Broadway, Ludgate, 1867.
- [A.M. Halsted]. *The Standard of Excellence as adopted by the American Poultry Society, being a reprint of the same as compiled and adopted by the London Poultry Club, with alterations and additions, adapting it to America*. New York: A.M. Halsted, 1867.
- William Bernhard Tegetmeier, Harrison Weir (illustrator). *The Poultry Book: comprising the breeding and management of profitable and ornamental poultry, their qualities and characteristics; to which is added "The Standard of Excellence in Exhibition Birds," authorized by the Poultry Club*. New edition, greatly enlarged. London; New York : George Routledge and Sons, the Broad-

way, Ludgate, 1873.

- ——. *The standard of excellence in exhibition poultry, authorised by the Poultry Club, to which is added the American standard. Reprinted from the original editions, with additions*. London, 1874.
- Alexander Comyns (editor). *The Standard of Perfection for Exhibition Poultry*. London: Office of the Poultry Club, [1886].
- T. Threlford (editor), Lewis Wright (introduction). *The Poultry Club Standards, containing a complete description of all the recognised varieties of fowls*, second edition. London; Paris; New York; Melbourne: Cassell & Company, 1901.
- Lewis Wright. *The New Book of Poultry. With Forty-Five Plates, by J.W. Ludlow, and the Poultry Club Standards of Perfection for the Various Breeds*. London; Paris; New York; Melbourne: Cassell and Company, 1902. (reprinted 1904)
- T. Threlford (editor), Lewis Wright (introduction). *The Poultry Club Standards. Containing a complete description of all the recognised varieties of fowls*, third edition. London: Cassell, 1905.
- Sidney Hubert Lewer, Lewis Wright. *Wright's Book of Poultry. Revised and edited in accordance with the latest Poultry Club standards by S. H. Lewer*. London; Paris; New York; Melbourne: Cassell and Company, [n.d., after 1910].
- William White Broomhead (editor). *The Poultry Club Standards: containing a complete description of all the recognised varieties of fowls, ducks, geese and turkeys*, fifth edition. London: The Poultry Club, 1922.
- ——. *The Poultry Club Standards*, sixth edition. London: Poultry Club of Great Britain, 1923.
- ——. *The Poultry Club Standards*, seventh edition. London: The Poultry Club, 1926.
- ——. *The Poultry Club Standards*, eighth edition. London: The Poultry Club, 1930.
- [Poultry Club of Great Britain]. *British Poultry Standards: Complete specifications and judging points of all standardised breeds and varieties of poultry as compiled by the specialist breed societies and recognised by The Poultry Club of Great Britain*. London: Poultry World in association with the Poultry Club of Great Britain, [1954].
- [Poultry Club of Great Britain]. *British Poultry Standards. Complete specifications and judging points of all standardised breeds and varieties of poultry as*

*compiled by the specialist breed societies and recognised by the Poultry Club of Great Britain*, revised edition, with illustrations. London: Poultry World, [1960].

- C. George May (editor). *British Poultry Standards: Complete specifications and judging points of all standardized breeds and varieties of poultry as compiled by the Specialist Breed Societies and recognized by the Poultry Club of Great Britain*. London: Iliffe Books, 1971.
- ——, David Hawksworth. *British Poultry Standards: Complete specifications and judging points of all standardized breeds and varieties of poultry as compiled by the Specialist Breed Societies and recognized by the Poultry Club of Great Britain*. London; Boston: Butterworth Scientific, 1982.
- ——. *British Poultry Standards: Complete specifications and judging points of all standardized breeds and varieties of poultry as compiled by the Specialist Breed Societies and recognized by the Poultry Club of Great Britain*, fourth edition, revised by David Hawksworth. Oxford: Blackwell Scientific, 1994. ISBN 0632038799.
- Victoria Roberts (editor). *British Poultry Standards: Complete specifications and judging points of all standardized breeds and varieties of poultry as compiled by the Specialist Breed Societies and recognized by the Poultry Club of Great Britain*, fifth edition. Oxford: Blackwell Science, 1997. ISBN 0632040521.
- ——. *British Poultry Standards: Complete specifications and judging points of all standardized breeds and varieties of poultry as compiled by the Specialist Breed Clubs and recognised by the Poultry Club of Great Britain*. Oxford: Blackwell, 2008. ISBN 9781405156424.

## References

- Ekarius, Carol (2007). Storey's Illustrated Guide to Poultry Breeds. 210 MAS MoCA Way, North Adams MA 01247: Storey Publishing. ISBN 978-1-58017-667-5.
- Heinricks, Christine (2007). How To Raise Chickens. Suite 200, 380 Jackson St, St Paul MN 55101: Voyageur Press. ISBN 978-0-7603-2828-6.
- Percy, Pam (2002). The Complete Chicken. Suite 200, 380 Jackson St, St Paul MN 55101: Voyageur Press. ISBN 0-89658-731-2.
- Staples, Tamara (2001). The Fairest Fowl. 85 Second Street San Francisco CA 94105: Chronicle Books LLC. ISBN 0-8118-3137-X.
- Victorian Poultry Fanciers Association; Poultry Stud Breeders and Exhibitors Victoria (2011), Australian poultry standards (2nd ed.), Victorian Poultry Fanciers Association Limited, ISBN 978-1-921488-23-8
- Hadlington, James; New South Wales. Dept. of Agriculture (1917), Poultry farming in New South Wales (2nd and rev. ed.), W.A. Gullick, Govt. Printer, retrieved 6 December 2013

# Permissions

All chapters in this book are published with permission under the Creative Commons Attribution Share Alike License or equivalent. Every chapter published in this book has been scrutinized by our experts. Their significance has been extensively debated. The topics covered herein carry significant information for a comprehensive understanding. They may even be implemented as practical applications or may be referred to as a beginning point for further studies.

We would like to thank the editorial team for lending their expertise to make the book truly unique. They have played a crucial role in the development of this book. Without their invaluable contributions this book wouldn't have been possible. They have made vital efforts to compile up to date information on the varied aspects of this subject to make this book a valuable addition to the collection of many professionals and students.

This book was conceptualized with the vision of imparting up-to-date and integrated information in this field. To ensure the same, a matchless editorial board was set up. Every individual on the board went through rigorous rounds of assessment to prove their worth. After which they invested a large part of their time researching and compiling the most relevant data for our readers.

The editorial board has been involved in producing this book since its inception. They have spent rigorous hours researching and exploring the diverse topics which have resulted in the successful publishing of this book. They have passed on their knowledge of decades through this book. To expedite this challenging task, the publisher supported the team at every step. A small team of assistant editors was also appointed to further simplify the editing procedure and attain best results for the readers.

Apart from the editorial board, the designing team has also invested a significant amount of their time in understanding the subject and creating the most relevant covers. They scrutinized every image to scout for the most suitable representation of the subject and create an appropriate cover for the book.

The publishing team has been an ardent support to the editorial, designing and production team. Their endless efforts to recruit the best for this project, has resulted in the accomplishment of this book. They are a veteran in the field of academics and their pool of knowledge is as vast as their experience in printing. Their expertise and guidance has proved useful at every step. Their uncompromising quality standards have made this book an exceptional effort. Their encouragement from time to time has been an inspiration for everyone.

The publisher and the editorial board hope that this book will prove to be a valuable piece of knowledge for students, practitioners and scholars across the globe.

# Index